SpringerBriefs in Applied Sciences and Technology

For further volumes:
http://www.springer.com/series/8884

Saptarshi Das · Indranil Pan

Fractional Order Signal Processing

Introductory Concepts and Applications

 Springer

Saptarshi Das
Department of Power Engineering
School of Nuclear Studies and
  Applications
Jadavpur University
Salt Lake Campus, LB-8, Sector 3
Kolkata, West Bengal 700098
India
e-mail: saptarshi@pe.jusl.ac.in;
  imsaptarshi@gmail.com

Indranil Pan
Department of Power Engineering
Jadavpur University
Salt Lake Campus, LB-8, Sector 3
Kolkata, West Bengal 700098
India
e-mail: indranil.jj@student.iitd.ac.in;
  indranil@pe.jusl.ac.in

ISSN 2191-530X
ISBN 978-3-642-23116-2
DOI 10.1007/978-3-642-23117-9
Springer Heidelberg Dordrecht London New York

e-ISSN 2191-5318
e-ISBN 978-3-642-23117-9

Cover design: eStudio Calamar, Berlin/Figueres

Printed on acid-free paper

Springer is part of Springer Science+Business Media (www.springer.com)

This book is dedicated to our beloved
Gurudev and Maa Tara

brahmarpanam brahma havir
brahmagnau brahmana hutam
brahmaiva tena gantavyam
brahma-karma-samadhina

—Jnana Yoga (Ch. 4), Verse 24,
Srimad Bhagavad Gita

Preface

Fractional Order Calculus has existed for over 300 years as an abstract mathematical concept. A renewed interest among the research community has come in the last few decades as fractional calculus has found applications in numerous science and engineering disciplines. Signal processing and control systems are two broad disciplines where fractional calculus has percolated to a substantial extent. Though there are a few books on fractional order calculus and fractional control systems, there are none on Fractional Order Signal Processing (FOSP). The idea of writing this book is to serve as a starting point for researchers interested in FOSP techniques.

Conventional Digital Signal Processing (DSP) deals with rational pole-zero models of FIR/IIR structures and looks at stability, causality, etc. of these models. Many physical signals and other phenomena have been shown to possess inherently fractional order dynamics and hence fractional calculus is naturally able to model these processes with greater accuracy. Hence we believe that fractional order signal and system theory will play a pivotal role in the coming eras of scientific development.

Theory of Statistical Signal Processing developed in parallel with DSP to analyze and handle signals with stochastic disturbances. However the major contributions in the field of Adaptive and Statistical Signal Processing have implicitly assumed the underlying probability distribution of the random variables to be Gaussian. Though this assumption works for some simplified cases, it does not deal with all possible real world scenarios where the distribution is inherently non-Gaussian. Among the many non-Gaussian random variables, a family of stable distributions which are capable of modeling signals with bursty nature have been found to effectively model many real world phenomena. These stable distributions can also be modeled from the perspective of fractional calculus and serve as the building blocks of fractional order Statistical Signal Processing.

Fourier transform and conventional filtering in Laplace (s) or z domain are widely used techniques in DSP. Fractional Fourier Transform (FrFT) extends this notion of frequency domain analysis to a more generalized time–frequency representation of the signal and can be used to separate the undesired component from the desired signal, even in cases where the integer order Fourier transform techniques fail. Classical Fourier Transform finds application in analyzing the desired and undesired components of a corrupted signal and the frequency domain information of a signal is usually used to design analog or digital filters. The flexibility of FrFT of any arbitrary order and filters with fractional pole-zero model together have opened new horizons in filter design theory.

Most of the literatures on FOSP are scattered in different research articles, technical reports and monographs. This book tries to assimilate these diverse topics and briefly present these concepts to the uninitiated researcher. The focus is on trying to cover a broad range of topics in FOSP and hence detailed discussion regarding each individual topic is out of scope of this book. However, this book can serve as a starting point and the researcher can delve into his topic of interest by following up the extensive list of references that have been included at the end of each chapter. An outline of the contents of the individual chapters is described next.

Chapter 1 gives a brief introduction to Generalized Non-integer Order Calculus. The basic definitions and useful realizations are presented briefly along with short explanations as to the physical interpretations of the same. The necessity of the application of fractional calculus in signal processing domain is next highlighted and probable application areas and current research focus in this domain is presented.

Chapter 2 looks at some Linear Time Invariant (LTI) representations of fractional order systems. Since the LTI theory is well developed, casting fractional order systems in state space or transfer function framework helps in the intuitive understanding and also the use of existing tools of LTI systems with some modifications. The discrete versions of fractional order models are also introduced since many practical applications in signal processing work on discrete time system models and discrete signals. Some basic signal processing techniques like continuous and discrete fractional order realizations, convolution and norms applied to fractional order signals and systems are also elucidated.

Chapter 3 explains long range dependent models which are useful in modeling real world signals having outliers or long tailed statistical distributions. The concept of stable distribution is outlined and many standard distributions like symmetric α-stable, Gaussian, Cauchy, etc. are obtained from the generalized description of the family of stable distributions. Self similarity for random processes is covered next and few other concepts like fractional Brownian motion, fractional and multi-fractional Gaussian noise, etc. are discussed in this context. The Hurst parameter which is a measure of self similarity is delineated next and a few methods to estimate it, is subsequently outlined.

Chapter 4 presents different transforms commonly encountered in signal processing applications. The importance of the fractional Fourier transform (FrFT) is highlighted and its interpretation and relation to various other transforms are also illustrated. The concept of filtering in the fractional Fourier domain and various other applications of signal processing using FrFT are also described. Other transforms like the fractional Sine, Cosine and the Hartley transforms which are obtained by suitable modification of the FrFT kernel is discussed. The fractional BSplines and fractional Wavelet transforms are also briefly introduced.

Chapter 5 explains various time domain and frequency domain system identification methods for fractional order systems from practical test data. System identification is important in cases where it is difficult to obtain the model from basic governing equations and first principles, or where there is only input–output data available and the underlying phenomenon is largely unknown. As is evident, fractional order models are better capable of modeling system dynamics than their integer order counterparts and hence identification using fractional order models is of practical interest from the system designer's point of view.

The first part of Chap. 6 throws light on the different forms of Kalman filter for fractional order systems. The latter part deals with signal processing with fractional lower order moments for α-stable processes. Some methods of parameter estimation of different α-stable processes are proposed. The concept of covariation, which is analogous to covariance for Gaussian signal processing, is unveiled and its properties and methods of estimation are reviewed.

Chapter 7 mentions several different MATLAB based toolboxes and codes which can be used to simulate fractional order signals and systems in general. Most of the toolboxes are in public domain and are freely accessible by any user. The reader can use these tools for simulation right away to get a feel of fractional order signal processing. These tools can also serve as building blocks for simulating more complex systems during research.

The authors would like to acknowledge the long technical discussions on FOSP with Shantanu Das (Scientist-H, Reactor Control Division, Bhabha Atomic Research Centre, Mumbai, India) and Basudev Majumdar (School of Nuclear Studies and Applications, Jadavpur University, Kolkata, India). The authors are also indebted to Prof. Dr. Amitava Gupta (Department of Power Engineering, Jadavpur University, Kolkata, India) for his encouragement and support. The authors are especially grateful to the growing community of researchers on FOSP who have made fundamental contributions to this topic and are responsible for opening myriad of new horizons in this field. Last but not the least; Saptarshi would like to acknowledge the inspiration received from his parents Supriya Das and Arunima Das and his brother Subhajit Das without whom the book would not have been a reality. Indranil would like to acknowledge the active participation of his father Prof. Dr. Tapan Kumar Pan in proof reading the manuscript and the loving understanding of his mother Sefali Pan throughout the countless hours spent working on the manuscript.

Saptarshi Das
Indranil Pan

Contents

Chapter 1
Introduction

Abstract This chapter gives a brief introduction to generalized non-integer order calculus. The basic definitions and useful realizations are presented briefly along with short explanations as to the physical interpretations of the same. The necessity of the application of fractional calculus in signal processing domain is next highlighted and probable application areas and current research focus in this domain is presented.

Keywords Fractional calculus · Mittag-Leffler · Grunwald-Letnikov · Riemann-Liouville

1.1 Basic Concepts of Fractional Calculus

Fractional calculus is the generalization of our familiar integer order successive differentiation and integration of any arbitrary order. Strictly speaking the term "fractional calculus" should be called non-integer order calculus as it essentially implies successive differ-integration of even irrational and complex order. It is 300 years old topic and has taken birth with the reply to a letter of Leibniz when he had been asked by L'Hospital what would be the result if the order of successive differentiation be 1/2. Leibniz replied that it would lead to an apparent paradox from which one day useful consequences would be drawn. The fractional calculus or generalized non-integer order calculus has gone through several modifications and changes to take a suitable form for being applied as the scientists' and engineers' tools. Early contributions to the theory of fractional calculus are made by many other renowned mathematicians. A detailed review on the history of fractional calculus can be found in Machado et al. (2011).

1.1.1 Special Functions

The generalized fractional calculus has been defined in various ways. The definition of fractional calculus uses the following three special functions:

S. Das and I. Pan, *Fractional Order Signal Processing*,
SpringerBriefs in Applied Sciences and Technology,
DOI: 10.1007/978-3-642-23117-9_1, © The Author(s) 2012

1.1.1.1 Gamma Function

The gamma function is the extension of the factorial for non-integer numbers.

$$\Gamma(z) := \int_0^\infty e^{-u} u^{z-1} du \quad \forall z \in \mathbb{R} \tag{1.1}$$

For complex values of z, the real part has to be positive to get finite value of the gamma function. The gamma function has the following property:

$$\Gamma(z+1) = z\Gamma(z) \tag{1.2}$$

1.1.1.2 Beta Function

The beta function is defined as:

$$B(p,q) := \int_0^1 (1-u)^{p-1} u^{q-1} du, \quad p, q \in \mathbb{R}_+ \tag{1.3}$$

The relation between beta and gamma function is given by the following relation.

$$B(p,q) = \frac{\Gamma(p)\Gamma(q)}{\Gamma(p+q)} = B(q,p), \quad p, q \in \mathbb{R}_+ \tag{1.4}$$

1.1.1.3 Mittag-Leffler Function

It is widely used as the solution of fractional order differential equations. It is basically a generalized higher transcendental which encompasses a wide variety of commonly encountered transcendental functions. The one parameter Mittag-Leffler function is defined as:

$$E_\alpha(z) := \sum_{k=0}^\infty \frac{z^k}{\Gamma(\alpha k + 1)}, \quad \alpha > 0 \tag{1.5}$$

The Mittag-Leffler function reduces to exponential function for $\alpha = 1$. The two parameter Mittag-Leffler function is defined as:

$$E_{\alpha,\beta}(z) := \sum_{k=0}^\infty \frac{z^k}{\Gamma(\alpha k + \beta)}, \quad \alpha > 0, \quad \beta > 0 \tag{1.6}$$

Clearly,

$$E_{\alpha,1}(z) = \sum_{k=0}^{\infty} \frac{z^k}{\Gamma(\alpha k + 1)} = E_{\alpha}(z) \tag{1.7}$$

Few special cases of Mittag-Leffler function yields some commonly used transcendental functions. In simplest case, Mittag-Leffler function takes the form of the exponential function and hence, Mittag-Leffler can be viewed as the generalized transcendental function e.g., variation in the second parameter in Eq. 1.6 yields:

$$E_{1,1}(z) = \sum_{k=0}^{\infty} \frac{z^k}{\Gamma(k + 1)} = \sum_{k=0}^{\infty} \frac{z^k}{k!} = e^z \tag{1.8}$$

$$E_{1,2}(z) = \sum_{k=0}^{\infty} \frac{z^k}{\Gamma(k + 2)} = \sum_{k=0}^{\infty} \frac{z^k}{(k + 1)!} = \frac{1}{z} \sum_{k=0}^{\infty} \frac{z^{k+1}}{(k + 1)!} = \frac{e^z - 1}{z} \tag{1.9}$$

$$E_{1,3}(z) = \sum_{k=0}^{\infty} \frac{z^k}{\Gamma(k + 3)} = \sum_{k=0}^{\infty} \frac{z^k}{(k + 2)!}$$
$$= \frac{1}{z^2} \sum_{k=0}^{\infty} \frac{z^{k+2}}{(k + 2)!} = \frac{e^z - 1 - z}{z^2} \tag{1.10}$$

As a generalized case,

$$E_{1,m}(z) = \frac{1}{z^{m-1}} \left(e^z - \sum_{k=0}^{m-2} \frac{z^k}{k!} \right) \tag{1.11}$$

Also, variation in the first parameter of Eq. 1.6 yields

$$E_{2,1}(z) = \sum_{k=0}^{\infty} \frac{z^{2k}}{\Gamma(2k + 1)} = \sum_{k=0}^{\infty} \frac{z^{2k}}{(2k)!} = \cosh(z) \tag{1.12}$$

$$E_{2,2}(z) = \sum_{k=0}^{\infty} \frac{z^{2k}}{\Gamma(2k + 2)} = \frac{1}{z} \sum_{k=0}^{\infty} \frac{z^{2k+1}}{(2k + 1)!} = \frac{\sinh(z)}{z} \tag{1.13}$$

1.1.2 Definitions in Fractional Calculus

The generalized fractional differentiation and integration has mainly two definitions as follows:

1.1.2.1 Grunwald-Letnikov (G-L) Definition

This formula is basically an extension of the backward finite difference formula for successive differentiation. This formula is widely used for the numerical solution of fractional differentiation or integration of a function. By Grunwald-Letnikov method the αth order differ-integration of a function $f(t)$ is defined as:

$$D_t^\alpha f(t) := \lim_{h \to 0} \frac{1}{h^\alpha} \sum_{j=0}^{\infty} (-1)^j \binom{\alpha}{j} f(t - jh) \qquad (1.14)$$

where

$$\binom{\alpha}{j} = \frac{\alpha!}{j!\,(\alpha - j)!} = \frac{\Gamma(\alpha + 1)}{\Gamma(j + 1)\,\Gamma(\alpha - j + 1)}$$

denotes the binomial co-efficients.

The Laplace transform of Grunwald-Letnikov fractional differ-integration is

$$\int_0^\infty e^{-st} {}_0D_t^\alpha f(t)dt = s^\alpha F(s) \qquad (1.15)$$

1.1.2.2 Riemann-Liouville (R-L) Definition

This definition is an extension of n-fold successive integration and is widely used for analytically finding fractional differ-integrals. By the Riemann-Liouville formula the αth order integration of a function $f(t)$ is defined as:

$$_aI_t^\alpha f(t) = {}_aD_t^{-\alpha} f(t) := \frac{1}{\Gamma(-\alpha)} \int_a^t \frac{f(\tau)}{(t - \tau)^{\alpha+1}} d\tau \qquad (1.16)$$

for $a, \alpha \in \mathbb{R}, \alpha < 0$.

By this formula fractional order differentiation is defined as the integer order successive differentiation of a fractional order integral. i.e.

$$_aD_t^\alpha f(t) := \frac{1}{\Gamma(n - \alpha)} \frac{d^n}{dt^n} \int_a^t \frac{f(\tau)}{(t - \tau)^{\alpha+1}} d\tau \qquad (1.17)$$

where $n - 1 < \alpha < n$.

The Laplace transform of Riemann-Liouville fractional differ-integration is:

$$\int_0^\infty e^{-st} {}_0D_t^\alpha f(t)dt = s^\alpha F(s) - \sum_{k=0}^{n-1} s^k \left. {}_0D_t^{\alpha-k-1} f(t) \right|_{t=0} \qquad (1.18)$$

1.2 Solution of Fractional Differential Equations Using Laplace Transform

If it be considered a physical system is governed by the following fractional differential equation:

$$aD_t^\alpha y(t) = f(t) \qquad (1.19)$$

with zero initial condition, the Laplace transform of the above equation yields

$$G_1(s) = \frac{Y(s)}{U(s)} = \frac{1}{as^\alpha} \qquad (1.20)$$

Inverse Laplace transform of the above equation produces

$$g_1(t) = \frac{1}{a}\frac{t^{\alpha-1}}{\Gamma(\alpha)} \qquad (1.21)$$

The above equation can be viewed as the impulse response of the system's transfer function model $G_1(s)$. Extending the concept for two term differential equation having the structure

$$aD_t^\alpha y(t) + by(t) = f(t) \qquad (1.22)$$

similar Laplace operation with zero initial condition gives the system's transfer function as

$$G_2(s) = \frac{Y(s)}{U(s)} = \frac{1}{as^\alpha + b} = \frac{1}{a} \cdot \frac{1}{s^\alpha + \frac{b}{a}} \qquad (1.23)$$

Inverse Laplace (Sheng et al. 2011) of the above equation gives

$$g_2(t) = \frac{1}{a}t^{\alpha-1}E_{\alpha,\alpha}\left(-\frac{b}{a}t^\alpha\right) \qquad (1.24)$$

The generalized solution of n–term fractional differential equation and other details can be found in Podlubny (1999).

1.3 Some Visualizations of Fractional Calculus

Impulse, step and ramp functions are most common in signals and systems theory because they have a generalized Laplace domain representation i.e.,

$$G(s) = \frac{1}{s^q} \qquad (1.25)$$

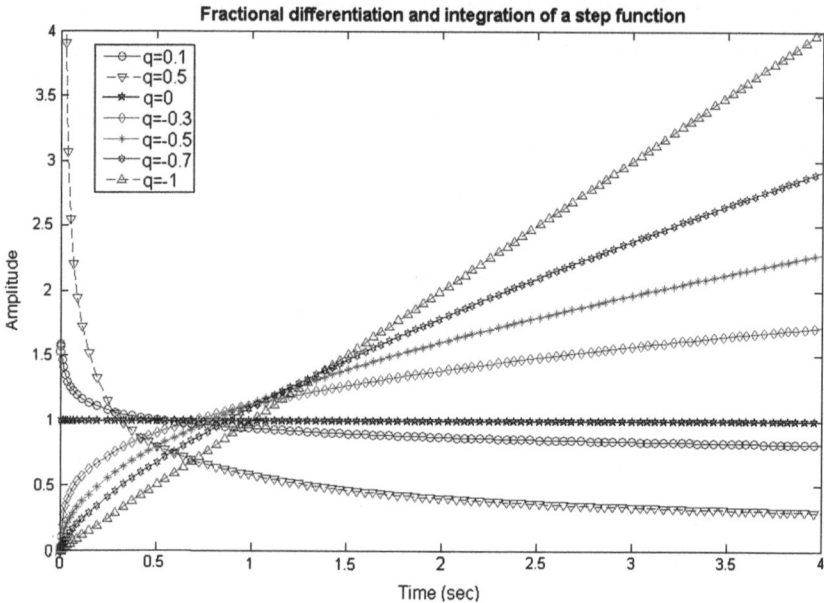

Fig. 1.1 Fractional differ-integration of a step function

where $q \in \{0, 1, 2\}$ for the respective cases.

It is known that the successive integer order integration of impulse function gives step and ramp, and the successive integer order differentiation of ramp function gives step and impulse functions. In Laplace domain, the impulse, step and ramp represent the unit delta operator, integrator and double-integrator, respectively. This idea can be extended to the generalized case where $q \in \mathbb{R}$ for any arbitrary order integration or differentiation. Figure 1.1 shows the time domain representation of different order of differ-integration of a step function. As can be seen from the figure, the time domain representations of the fractional order cases lay between the discrete integer order operations.

As it is known from the system theory, a physical system governed by second order differential equation of the form

$$\frac{d^2 y(t)}{dt^2} + 2\xi\omega\frac{dy(t)}{dt} + \omega^2 = u(t) \tag{1.26}$$

has oscillatory time response for $\xi \in (0, 1)$ and sluggish time response for $\xi > 1$. Here $\{\xi, \omega\}$ signifies the system's damping ratio and natural frequency, respectively. For impulse input excitation in (1.26), the oscillations decay with an exponential envelope.

$$\mathcal{L}^{-1}\frac{Y(s)}{U(s)} = \frac{e^{-\xi\omega t}}{\omega\sqrt{1 - \xi^2}} \sin\left(\omega t\sqrt{1 - \xi^2}\right) \tag{1.27}$$

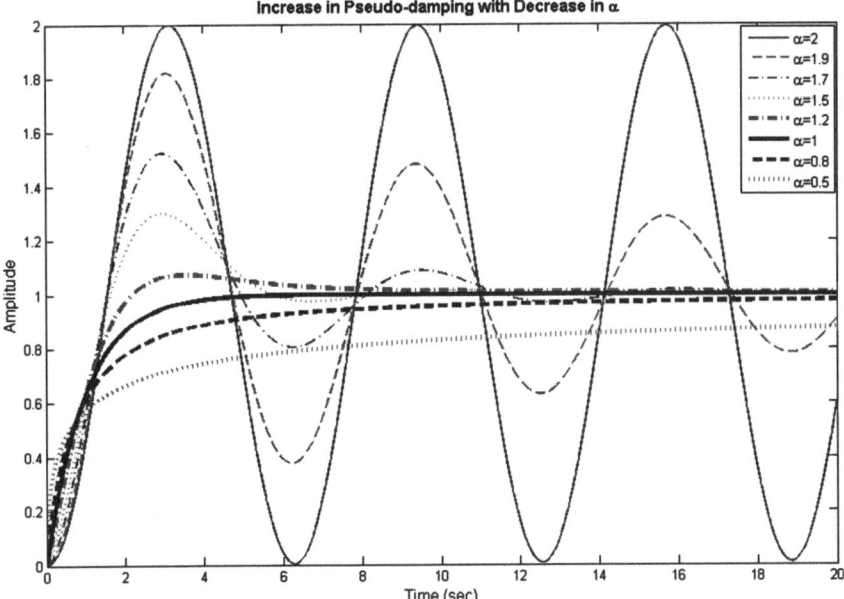

Fig. 1.2 Increase in pseudo-damping with decrease in fractional order of the system

But for fractional order system, governed by a differential equation whose maximum order of differentiation is a fraction e.g., (1.22), the impulse response also decays in a oscillatory manner though there is no explicit damping term in the governing differential equation. This phenomenon can be termed as the pseudo-damping which becomes predominant with decrease in the order of fractional differential equation and the oscillation decays with a power law as in Eq. 1.24. Putting $a = b = 1$, in Eq. 1.23, the step response is plotted by varying the order α in Fig. 1.2. The figure shows that the decrease in order α results in an increase in the pseudo-damping.

1.4 Geometric and Physical Interpretation of Fractional Order Differentiation and Integration

The geometric interpretation of fractional order (FO) differ-integrations are more complex than their integer order counterparts and have intrigued researchers for years. An interpretation according to Podlubny (2002) is given below.

The left-sided Riemann-Liouville fractional integral of order α is given as

$$_0I_t^\alpha f(t) = \frac{1}{\Gamma(\alpha)} \int_0^t f(\tau)(t - \tau)^{\alpha-1}d\tau, \quad t \geq 0 \tag{1.28}$$

The time variable τ in Eq. 1.28 can be replaced by a scale transformation $g_t(\tau)$, i.e. $\tau \to g_t(\tau)$. Taking

$$g_t(\tau) = \frac{1}{\Gamma(\alpha+1)} \left\{ t^\alpha - (t-\tau)^\alpha \right\} \tag{1.29}$$

we have,

$$dg_t(\tau) = \frac{(t-\tau)^{\alpha-1}}{\Gamma(\alpha)} \tag{1.30}$$

Hence, (1.28) can be written in the form

$${}_0I_t^\alpha f(t) = \int_0^t f(\tau) dg_t(\tau) \tag{1.31}$$

Keeping t fixed, let us consider a 3D curve in the space (τ, g, f) given by the following expression

$$C_t : (\tau, g_t(\tau), f(\tau)), \quad 0 \leq \tau \leq t \tag{1.32}$$

Along the curve C_t if a "fence" is built perpendicular to the plane (τ, g) of varying height $f(\tau)$ as in Fig. 1.3, then the shadows cast on the walls by the fence may be interpreted as follows.

1. The area of the projection of the fence on the plane (τ, f) is given by

$$I_0^1 f(t) = \int_0^t f(\tau) d\tau \tag{1.33}$$

2. The area of the projection of the fence on the plane (g, f) is given by

$$I_0^\alpha f(t) = \int_0^t f(\tau) dg(\tau) \tag{1.34}$$

For $\alpha = 1$, Eq. 1.29 reduces to $g_t(\tau) = \tau$ and hence both the shadows are equal. Thus the integer order definite integration is a special case of the left sided Riemann-Liouville fractional integration even from a geometric perspective. When t is changing, the fence changes in length and shape. The corresponding changes in the shadow on the walls (g, f) (as shown in Fig. 1.4) due to the change of the fence with time give a dynamical geometric interpretation of the fractional integral given by Eq. 1.28 as a function of the variable t.

The physical interpretation of $I_0^\alpha f$ can be obtained by introducing the notion of a transformed time scale where the time does not flow homogeneously. Considering

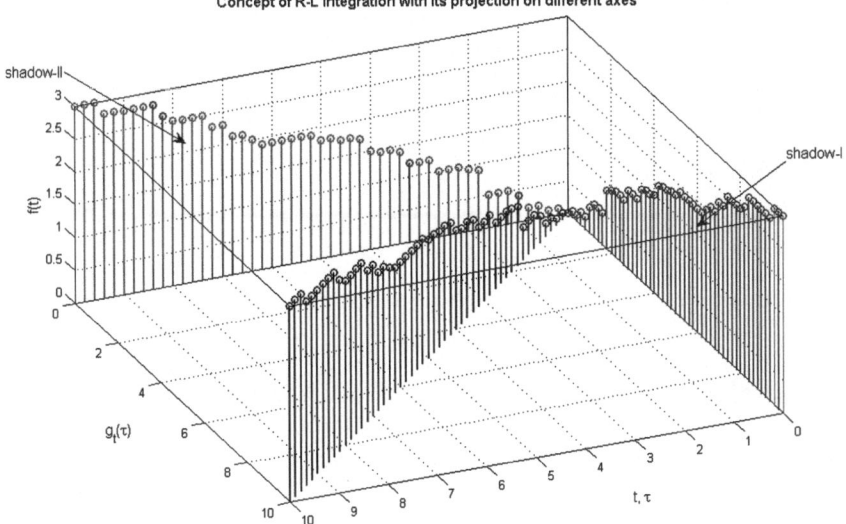

Concept of R-L integration with its projection on different axes

Fig. 1.3 Geometric interpretation of fractional order integration

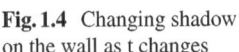

Fig. 1.4 Changing shadow
on the wall as t changes

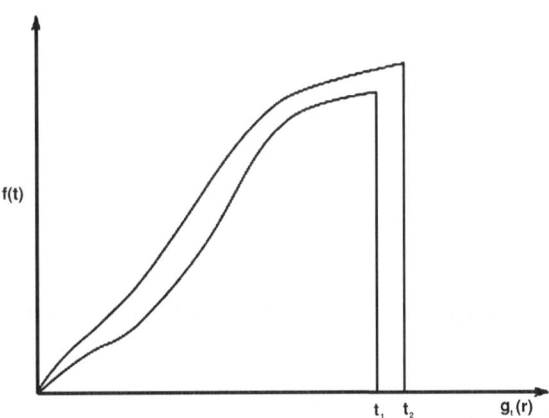

τ as the time, the third dimension $g_t(\tau)$ added to the pair $(\tau, f(\tau))$ can be considered as some kind of a transformed time scale. Thus arises the concept of two kinds of time as represented in Fig. 1.5.

1. The mathematical time τ which is assumed to be homogeneous and equably flowing.
2. The transformed time $g(\tau)$ whose notion can be understood from the following. Assuming a clock displays the time τ incorrectly and the relationship between the measured time τ and the real time T (i.e. the correct or transformed time) is given by $T = g(\tau)$.

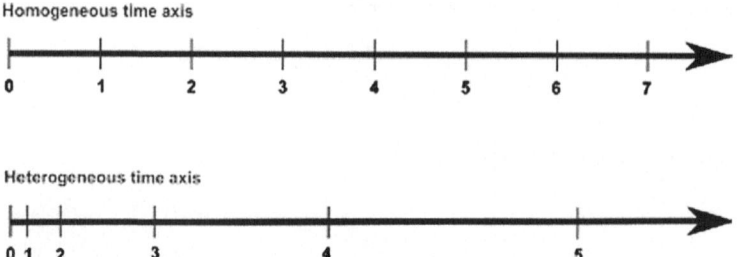

Fig. 1.5 The concept of homogeneous and heterogeneous time

Hence when a real time interval of $dT = dg(\tau)$ elapses, the time interval measured using the notion of mathematical time is $d\tau$. Thus if $v(\tau)$ is the measured velocity of a body, then the wrong value of distance covered is given by the integral

$$I_0^1 v(t) = \int_0^t v(\tau)d\tau \qquad (1.35)$$

whereas the real or actual distance passed is given by

$$I_0^\alpha v(t) = \int_0^t v(\tau)dg(\tau) \qquad (1.36)$$

1.5 Application of Fractional Calculus in Science and Engineering

The application of fractional calculus in the enhancement of systems and control theory has already been popular amongst the research community. Several books have been written by the contemporary researchers like Podlubny (1999), Miller and Ross (1993), Oldham and Spanier (1974), Kilbas et al. (2006), Das (2008), Diethelm (2010), Hilfer (2000), Ortigueira (2011), Caponetto et al. (2010) to make the subject well understandable for a practicing engineer's or scientist's point of view. This book presents few well-established signal processing techniques in the light of fractional order calculus.

Three special issues in Elsevier's peer reviewed journal (Signal Processing) have been dedicated to the advancement of fractional calculus in signal processing (Ortigueira and Machado 2003, 2006; Ortigueira et al. 2011). The various applications of fractional order signal processing (FOSP) includes, chemistry, biology, economics, control theory, signal and image processing, thermal engineering, acoustics, electromagnetic theory, robotics, visco-elasticity, diffusion, turbulence, mechanics. The major sub-branches of FOSP are analysis of fractional order linear system theory, application of fractional Fourier transform in filtering, electrochemical realization of fractional impedance or fractance, modeling long memory processes or time-series, estimation and identification of fractional systems, etc.

Recent advances in fractional calculus and its engineering applications have been published in various literatures like journal special issues, conferences, edited and authored books. Extensively review on this domain can be found in Machado et al. (2011), Debnath (2003) and Gutierrez et al. (2010). Details of fractional calculus and its various applications can also be found in Baleanu et al. (2010), Oldham and Spanier (1974), Podlubny (1999) Sabatier et al. (2007), Miller and Ross (1993), Caponetto et al. (2010), Monje et al. (2010), Das (2008) and Mainardi (2009).

References

Baleanu, D., Guven, Z.B., Machado, J.A.T.: New Trends in Nanotechnology and Fractional Calculus Applications. Springer, Heidelberg (2010)

Caponetto, R., Dongola, G., Fortuna, L.: Fractional Order Systems: Modeling and Control Applications, vol. 72, World Scientific Publishing, Singapore (2010)

Das, S.: Functional Fractional Calculus for System Identification and Controls. Springer, Heidelberg (2008)

Debnath, L.: Recent applications of fractional calculus to science and engineering. J. Math. Math. Sci. **54**, 3413–3442 (2003)

Diethelm, K.: The Analysis of Fractional Differential Equations: An Application-Oriented Exposition Using Differential Operators of Caputo Type, vol. 2004, Springer, Heidelberg (2010)

Gutierrez, R.E., Rosario, J.M., Tenreiro Machado, J.: Fractional Order Calculus: Basic Concepts and Engineering Applications. Mathematical Problems in Engineering: Theory, Methods and Applications. Art. no. 375858, Hindawi Publishing Corporation (2010). doi: 10.1155/2010/375858

Hilfer, R.: Applications of Fractional Calculus in Physics. vol. 1600, World Scientific, Singapore (2000)

Kilbas, A.A., Srivastava, H.M., Trujillo, J.J.: Theory and Applications of Fractional Differential Equations, vol. 204. Elsevier, Amsterdam (2006)

Machado, J.T., Kiryakova, V., Mainardi, F.: Recent history of fractional calculus. Commun. Nonlinear Sci. Numer. Simul. **16**(3), 1140–1153 (2011). doi:10.1016/j.cnsns.2010.05.027

Mainardi, F.: Fractional Calculus and Waves in Linear Viscoelasticity: An Introduction to Mathematical Models. Imperial College Press, London (2009)

Miller, K., Ross, B.: An Introduction to the Fractional Calculus and Fractional Differential Equations. Wiley, New Yok (1993)

Monje, C.A., Chen, Y.Q., Vinagre, B.M., Xue, D., Feliu, V.: Fractional-Order Systems and Controls: Fundamentals and Applications. Springer, Heidelberg (2010)

Oldham, K.B., Spanier, J.: The Fractional Calculus. Academic Press, New York (1974)

Ortigueira, M.D.: Fractional Calculus for Scientists and Engineers, vol. 10. Springer, Heidelberg (2011)

Ortigueira, M.D., Machado, J.A.T.: Fractional signal processing and applications. Signal Process. **83**(11), 2285–2286 (2003). doi:10.1016/s0165-1684(03)00181-6

Ortigueira, M.D., Machado, J.A.T.: Fractional calculus applications in signals and systems. Signal Process. **86**(10), 2503–2504 (2006). doi:10.1016/j.sigpro.2006.02.001

Ortigueira, M.D., Machado, J.A.T., Trujillo, J.J., Vinagre, B.M.: Fractional signals and systems. Signal Process. **91**(3), 349–349 (2011). doi:10.1016/j.sigpro.2010.08.002

Podlubny, I.: Fractional Differential Equations. Mathematics in Science and Engineering, vol. 198. Academic Press, New York (1999)

Podlubny, I.: Geometric and physical interpretation of fractional integration and fractional differentiation. Fract. Calc. Appl. Anal. **5**(4) (2002)

Sabatier, J., Agrawal, O.P., Machado, J.A.T.: Advances in Fractional Calculus: Theoretical Developments and Applications in Physics and Engineering. Springer, Heidelberg (2007)

Sheng, H., Li, Y., Chen, Y.: Application of numerical inverse Laplace transform algorithms in fractional calculus. J. Frankl. Inst. **348**(2), 315–330 (2011). doi:10.1016/j.jfranklin.2010.11.009

Chapter 2
Basics of Fractional Order Signals and Systems

Abstract This chapter looks at some linear time invariant (LTI) representations of fractional order systems. Since the LTI theory is well developed, casting fractional order systems in state space or transfer function framework helps in the intuitive understanding and also the use of existing tools of LTI systems with some modifications. The discrete versions of fractional order models are also introduced since many practical applications in signal processing work on discrete time system models and discrete signals. Some basic signal processing techniques, like continuous and discrete fractional order realizations, convolution and norms, applied to fractional order signals and systems are also introduced.

Keywords Fractional order (FO) differ-integrators · Constant Phase Element (CPE) · FO realization · FO LTI systems · FO convolution · FO correlation

Modern science has tried to interpret nature from a mathematical framework and has been successful to a large extent in analyzing various phenomena and harnessing control over various physical processes. But on introspection we need to look at the very mathematical tools in our quest for perfection and understanding. Fractional order differential and integral equations are a generalization of our conventional integral and differential equations that extend our notion of modeling the real world around us. The term "fractional" basically implies all non-integer numbers (e.g. fractions, irrational numbers and even complex numbers) and hence a more mathematically correct nomenclature is non-integer order calculus. However, the present focus of the book is real non-integer numbers and the term 'fractional' used throughout this book is in this sense.

There are many processes in nature that can be more accurately modeled using fractional differ-integrals. It has been experimentally demonstrated that the charging and discharging of lossy capacitors for example, follows inherently fractional order dynamics. The flow of fluid in a porous media, the conduction of heat in a semi-infinite slab, voltage-current relation in a semi infinite transmission line, non-Fickian diffusion , etc. are all such examples where the governing equations can be modeled more accurately using fractional order differential or integral operators.

A Fractional order linear time invariant (LTI) system is mathematically equivalent to an infinite dimensional LTI filter. Thus a fractional order system can be approximated using higher order polynomials having integer order differ-integral operators. The concept can be viewed similar to the Taylor series expansion of non-linear functions as a summation of several linear weighted differentials. More involved analysis and review of existing fractional order signals and systems can be found in Magin et al. (2011), Ortigueira (2000a, b, 2003, 2006, 2008, 2010).

2.1 Fractional Order LTI Systems

Fractional order LTI systems can be modeled using the conventional input–output transfer function approach similar to the integer order ordinary differential equations (Monje et al. 2010). Fractional order ordinary differential equations lead to fractional order transfer function models by the well known Laplace transform technique. The conventional notion of state space modeling can also be extended to represent fractional order dynamical system models with additional requirement of defining the state variables as fractional derivative of another.

2.1.1 Transfer Function Representation

Let us consider the following fractional differential equation to represent the dynamics of a system:

$$a_n D^{\alpha_n} y(t) + a_{n-1} D^{\alpha_{n-1}} y(t) + \cdots + a_0 D^{\alpha_0} y(t)$$
$$= b_m D^{\beta_m} u(t) + b_{m-1} D^{\beta_{m-1}} u(t) + \cdots + b_0 D^{\beta_0} u(t) \qquad (2.1)$$

In the above fractional differential equation if the order of differentiations be integer multiple of a single base order i.e. $\alpha_k, \beta_k = k\alpha, \quad \alpha \in \mathbb{R}_+$, the system will be termed as commensurate order and takes the following form

$$\sum_{k=0}^{n} a_k D^{k\alpha} y(t) = \sum_{k=0}^{m} b_k D^{k\alpha} u(t) \qquad (2.2)$$

Taking Laplace transform of the above equation and putting zero initial condition, the input–output fractional order transfer function models takes the form

$$G(s) = \frac{Y(s)}{U(s)} = \frac{b_m s^{\beta_m} + b_{m-1} s^{\beta_{m-1}} + \cdots + b_0 s^{\beta_0}}{a_n s^{\alpha_n} + a_{n-1} s^{\alpha_{n-1}} + \cdots + a_0 s^{\alpha_0}} \qquad (2.3)$$

For commensurate fractional order systems the above transfer function takes the following form

$$G(s) = \frac{\sum_{k=0}^{m} b_k (s^\alpha)^k}{\sum_{k=0}^{n} a_k (s^\alpha)^k} = \frac{\sum_{k=0}^{m} b_k \lambda^k}{\sum_{k=0}^{n} a_k \lambda^k}, \quad \lambda = s^\alpha \qquad (2.4)$$

Fractional order transfer function models leads to the concept of fractional poles and zeros in the complex s-plane (Merrikh-Bayat et al. 2009). For the commensurate fractional order system (2.4) with the characteristic equation in terms of the complex variable λ is said to be bounded-input bounded-output (BIBO) stable, if the following condition is satisfied

$$|\arg(\lambda_i)| > \alpha \frac{\pi}{2} \qquad (2.5)$$

where λ_i are the roots of the characteristic polynomial in λ.

For $\lambda = 1$, the well known stability condition for integer order transfer functions is obtained which states the real part of the poles should lie in the negative half of the complex s-plane i.e. $|\arg(\lambda_i)| > \frac{\pi}{2}$.

2.1.2 State-Space Representation

A generalized fractional order multiple input multiple output (MIMO) LTI state-space model can be represented as

$$\begin{aligned} D^\alpha x &= Ax + Bu \\ y &= Cx + Du \end{aligned} \qquad (2.6)$$

where $\alpha = [\alpha_1, \alpha_2, \ldots, \alpha_n]$ is the commensurate or incommensurate fractional orders. $u \in \mathbb{R}^l$ is the input column vector, $x \in \mathbb{R}^n$ is the state column vector, $y \in \mathbb{R}^p$ is the output column vector, $A \in \mathbb{R}^{n \times n}$ is the state matrix, $B \in \mathbb{R}^{n \times l}$ is the input matrix, $C \in \mathbb{R}^{p \times n}$ is the output matrix, $D \in \mathbb{R}^{p \times l}$ is the direct transmission matrix. In the above fractional order state-space representation, the first equation is called the fractional order state equation and the second one is known as the output equation. The fractional state space model can be converted to the fractional order transfer function form using the following relation

$$G(s) = C \left(s^\alpha I - A\right)^{-1} B + D \qquad (2.7)$$

Here, I represents the identity matrix of dimension $n \times n$ and $G(s)$ represents the fractional order transfer function matrix of dimension $p \times l$. The fractional order state-space realization in controllable, observable and diagonal canonical forms are similar to the corresponding integer order state-space models and have been detailed in Monje et al. (2010). Accuracy of the numerical approximation of FO state-space has been detailed in Rachid et al. (2011) and Bouafoura et al. (2011).

It is interesting to note that the stability analysis of commensurate fractional order systems is relatively easier (Trigeassou et al. 2011; Trigeassou and Maamri 2011) since the conventional s-plane instability region gets contracted by $\alpha\pi/2$ for $0 < \alpha < 1$. Stability of incommensurate fractional order models are mathematically more involved and have been discussed by Matignon (1998) in a detailed manner.

The discrete time fractional order state space corresponding to the continuous time fractional order state space can be represented as

$$
\left.
\begin{aligned}
x\,(k+1) &= \left(AT^\alpha + \alpha I\right) x\,(k) + BT^\alpha u\,(k) && \text{for } k=0 \\
x\,(k+1) &= \left(AT^\alpha + \alpha I\right) x\,(k) \\
&\quad - \sum_{i=2}^{k+1} (-1)^i \binom{\alpha}{i} x\,(k+1-i) + BT^\alpha u\,(k) && \text{for } k \geq 1 \\
y\,(k) &= Cx\,(k) + Du\,(k)
\end{aligned}
\right\} \quad (2.8)
$$

Considering an infinite dimensional memory of a fractional order system, the above discrete time fractional order state space expressed in terms of an expanded state space

$$
\begin{bmatrix} x\,(k+1) \\ x\,(k) \\ x\,(k-1) \\ \vdots \end{bmatrix} = \tilde{A} \begin{bmatrix} x\,(k) \\ x\,(k-1) \\ x\,(k-2) \\ \vdots \end{bmatrix} + \tilde{B} u\,(k)
$$

$$
y\,(k) = \tilde{C} \begin{bmatrix} x\,(k) \\ x\,(k-1) \\ x\,(k-2) \\ \vdots \end{bmatrix} + Du\,(k)
$$

$$(2.9)$$

where

$$
\tilde{A} = \begin{bmatrix} (AT^\alpha + \alpha I) & -I\,(-1)^2 \binom{\alpha}{2} & -I\,(-1)^3 \binom{\alpha}{3} & \cdots \\ I & 0 & 0 & \cdots \\ 0 & I & 0 & \cdots \\ \vdots & \vdots & \vdots & \ddots \end{bmatrix}, \quad \tilde{B} = \begin{bmatrix} BT^\alpha \\ 0 \\ 0 \\ \vdots \end{bmatrix},
$$

$$
\tilde{C} = [C \quad 0 \quad 0 \quad \cdots]
$$

Here, 0 is the null matrix of dimension $n \times n$.

The above discrete time fractional order state space model is asymptotically stable if the following condition $\|\tilde{A}\| < 1$ is satisfied. Here, $\|\cdot\|$ denotes the matrix norm defined as $\max |\lambda_i|$ and λ_i being the ith eigen-value of extended matrix $\tilde{A}$.

2.2 Realization of Fractional Order Differ-integrators

The realization of FO differ-integrators can be done in two ways viz. continuous time and discrete time realization.

2.2.1 Continuous Time Realization

Fractional order elements generally shows a constant phase curve, hence are also known as constant phase elements (CPE). In practice, a fractional order element can be approximated as a higher order system which maintains a constant phase within a chosen frequency band (Krishna 2011; Elwakil 2010; Maundy et al. 2011). The fractional order elements can be rationalized by several iterative techniques namely, Carlson's method (Carlson and Halijak 1964), Oustaloup's method (Oustaloup et al. 2000), Charef's method (Charef et al. 1992), etc. some of which are detailed below.

2.2.1.1 Carlson's Method

Fractional order elements or transfer functions can be recursively approximated using the following formulation known as Carslon's method (Carlson and Halijak 1964).

If $G(s)$ be a rational transfer function and $H(s)$ be a fractional order transfer function such that $H(s) = [G(s)]^q$ where $q = m/p$ is a fractional order of the transfer function, then $H(s)$ can be recursively approximated as

$$H_i(s) = H_{i-1}(s) \frac{(p-m)\left[H_{i-1}(s)\right]^2 + (p+m)G(s)}{(p+m)\left[H_{i-1}(s)\right]^2 + (p-m)G(s)} \tag{2.10}$$

with an initial guess of $H_0(s) = 1$.

The above recursive formula can be rewritten as

$$H_i(s) = H_{i-1}(s) \frac{G(s) + \alpha \left[H_{i-1}(s)\right]^2}{\alpha G(s) + \left[H_{i-1}(s)\right]^2} \tag{2.11}$$

where $\alpha := \frac{p-m}{p+m} = \left[\frac{1-\left(\frac{m}{p}\right)}{1+\left(\frac{m}{p}\right)}\right] = \frac{1-q}{1+q}$ and $q := \frac{1-\alpha}{1+\alpha}$

Therefore, for the simplest case i.e. $H(s) = s^q$, it will act like a differentiator for $q > 0$ implying $\alpha < 1$. Also, it will act like an integrator for $q < 0$ implying $\alpha > 1$.

2.2.1.2 Charef's Method

Basic formulation of Charef's approximation technique is detailed in Charef et al. (1992).

Let us take a fractional order transfer function

$$H(s) = \frac{1}{(1 + s/p_T)^m} \quad (2.12)$$

where p_T is the pole of the FO system and m is the fractional order of the system. The above all pole fractional order system can be rationalized by the following recursive formula

$$H(s) = \frac{\prod_{i=0}^{N-1}\left(1 + \frac{s}{z_i}\right)}{\prod_{i=0}^{N}\left(1 + \frac{s}{p_i}\right)} \quad (2.13)$$

If the maximum allowable discrepancy in estimating the frequency response of the approximated model be $y\ dB$, the zeros and poles can be recursively calculated as

$$z_{N-1} = p_{N-1}10^{[y/10(1-m)]}$$
$$p_N = z_{N-1}10^{[y/10m]} \quad (2.14)$$

The first approximation of the pole and zero starts with

$$p_0 = p_T 10^{[y/20m]}$$
$$z_0 = p_0 10^{[y/10(1-m)]} \quad (2.15)$$

Now the problem is to determine the value of N so that a specified accuracy of the approximated rational transfer function at the corner frequencies can be obtained.

Let, $a = 10^{[y/10(1-m)]} = \dfrac{z_{N-1}}{p_{N-1}}$ and $b = 10^{[y/10m]} = \dfrac{p_N}{z_{N-1}}$ then

$ab = 10^{[y/10m(1-m)]} = \dfrac{z_{N-1}}{z_{N-2}} = \dfrac{p_N}{p_{N-1}}.$

Now, N can be determined by the following expression

$$N = Integer\left(\frac{\log\left(\frac{\omega_{max}}{p_0}\right)}{\log(ab)}\right) + 1 \quad (2.16)$$

2.2.1.3 Oustaloup's Method

Oustaloup's recursive filter gives a very good fitting to the fractional-order elements s^γ within a chosen frequency band (Oustaloup et al. 2000). Let us assume that the expected fitting range is (ω_b, ω_h). The filter can be written as:

$$G_f(s) = s^\gamma = K \prod_{k=-N}^{N} \frac{s + \omega_k'}{s + \omega_k} \quad (2.17)$$

where the poles, zeros, and gain of the filter can be evaluated as

$$\omega_k = \omega_b \left(\frac{\omega_h}{\omega_b}\right)^{\frac{k+N+\frac{1}{2}(1+\gamma)}{2N+1}}, \omega'_k = \omega_b \left(\frac{\omega_h}{\omega_b}\right)^{\frac{k+N+\frac{1}{2}(1-\gamma)}{2N+1}}, K = \omega_h^\gamma$$

2.2.2 Discrete Time Realization

2.2.2.1 Based on Discretization Method

Discrete time realizations are often preferred over continuous time realizations since they can be easily implemented in hardware and updated with the change in the purpose of application. There are mainly two methods of discretization viz. indirect and direct method. The indirect discretization method is accomplished in two steps. Firstly the frequency domain fitting is done in continuous time domain and then the fitted continuous time transfer function is discretized. Direct discretization based methods include the application of power series expansion (PSE), continued fraction expansion (CFE), MacLaurin series expansion, etc. with a suitable generating function. The mapping relation or formula for conversion from continuous time to discrete time operator ($s \leftrightarrow z$) is known as the generating function. Stability of direct discretization has been analyzed in Siami et al. (2011). Among the family of expansion methods, CFE based digital realization has been extensively studied with various types of generating functions like Tustin, Simpson, Al-Alaoui, mixed Tustin-Simpson, mixed Euler-Tustin-Simpson, impulse response based and other higher order generating functions (Gupta et al. 2011; Chen et al. 2004; Visweswaran et al. 2011).

When the continuous time realization like Oustaloup's approximation , etc. with a constant phase is discretized with a suitable generating function the phase of the realized filter may deviate from the original desired one depending on the generating function. However, a direct discretization scheme tries to maintain a constant phase of the FO element directly in the frequency domain. Hence, direct discretization is generally preferred for digital realization over its indirect counterpart.

- *Generating Functions*: The discrete time rational approximation of a simple continuous-time differentiator $s \approx H\left(z^{-1}\right)$ is as follows

  – *Euler (Rectangular Rule)*:

$$H_{Euler}\left(z^{-1}\right) = \left[\frac{1-z^{-1}}{T}\right] \tag{2.18}$$

Here, T represents the sampling time and z denotes the discrete time complex frequency. It is clear that the Euler's discretization formula (2.18) is an extension of the backward difference technique of numerical differentiation. Forward or central differencing schemes is generally not used since it would result in a non causal filter, i.e. future values of a series would be required to obtain the present value.

- Tustin (*Trapezoidal Rule or Bilinear Transform*):

$$H_{Tustin}\left(z^{-1}\right) = \left[\frac{2}{T} \cdot \frac{1-z^{-1}}{1+z^{-1}}\right] \tag{2.19}$$

The Tustin's discretization can be obtained from the basic $(s \leftrightarrow z)$ mapping relation by expanding the exponential terms with their first order approximations.

$$z = e^{sT} = \frac{e^{sT/2}}{e^{-sT/2}} = \left(1+\frac{sT}{2}\right)\Big/\left(1-\frac{sT}{2}\right) = \frac{2+sT}{2-sT}$$

$$\Rightarrow s = \frac{2}{T} \cdot \left(\frac{z-1}{z+1}\right) \tag{2.20}$$

- *Simpson's Rule*: The well known Simpson's numerical integration formula is given by (in time domain)

$$y(n) = \frac{T}{3}[x(n) + 4x(n-1) + x(n-2)] + y(n-2) \tag{2.21}$$

By applying z transform on (2.21) it is found

$$\frac{Y(z)}{X(z)} = H(z) = \frac{T}{3}\left(\frac{1+4z^{-1}+z^{-2}}{1-z^{-2}}\right) \tag{2.22}$$

The relation given by (2.22) represents a digital integrator and can be inverted to obtain a digital differentiator as

$$H_{Simpson}\left(z^{-1}\right) = \left[\frac{3}{T} \cdot \frac{\left(1+z^{-1}\right)\left(1-z^{-1}\right)}{1+4z^{-1}+z^{-2}}\right] \tag{2.23}$$

- *Al-Alaoui Operator*: Al-Alaoui has shown that the discretization formula can be improved by interpolating the classical Euler and Tustin's formula as follows

$$H_{Al\text{-}Alaoui(\alpha)}\left(z^{-1}\right) = \alpha H_{Euler}\left(z^{-1}\right) + (1-\alpha) H_{Tustin}\left(z^{-1}\right)$$

$$= \frac{\alpha T}{\left(1-z^{-1}\right)} + \frac{(1-\alpha)T}{2}\left(\frac{1+z^{-1}}{1-z^{-1}}\right) = \frac{T\left[(1+\alpha)+(1-\alpha)z^{-1}\right]}{2\left(1-z^{-1}\right)}$$

$$= \frac{T(1+\alpha)}{2} \cdot \frac{\left(1+\frac{(1-\alpha)}{(1+\alpha)}z^{-1}\right)}{\left(1-z^{-1}\right)}$$

$$\tag{2.24}$$

where $\alpha \in (0, 1)$ is a user-specified weight that balances the impact of the two generating function i.e. Euler (Rectangular) and Tustin (Trapezoidal) and their corresponding accuracies introduced in the discretization. Replacing $\alpha = 3/4$ in (2.24) produces the conventional Al-Alaoui operator as:

$$H_{Al-Alaoui(3/4)}\left(z^{-1}\right) = \frac{7T}{8} \cdot \frac{\left(1 + z^{-1}/7\right)}{\left(1 - z^{-1}\right)} \tag{2.25}$$

Generalized Al-Alaoui operator (2.24) shows that the Infinite Impulse Response (IIR) filter has a pole at $z = 1$ and zero varies between $z \in [-1, 0]$ for $\alpha \in [0, 1]$. Thus, the operator (2.24) can be directly inverted to produce a stable IIR realization for a differentiator also.

Early discretization techniques developed by Euler and Tustin are mainly based on the first order polynomial fitting. Simpson's advancement in the discretization technique shows that one can fit higher order polynomial to obtain better accuracy. But this is not a feasible proposition, since expansion with higher order generating function would increase the overall order of the discrete time filter. Also, as the order of the generating function increases, the region of performance in the frequency domain gets shrunk. The generalized Al-Alaoui type generating function (2.24) is ideal for applications where the requirement is to maximize accuracy without going for a higher order realization.

Other higher order generating functions like Chen-Vinagre operator (Chen and Vinagre 2003), Schneider operator, Al-Alaoui–Schneider–Kaneshige–Groutage (Al-Alaoui–SKG) operator, Hsue operator, Barbosa operator, Maione operator , etc. are also used. For details please look in Visweswaran et al. (2011), Gupta et al. (2011), Barbosa et al. (2006), Maione (2008) and the references therein.

2.2.2.2 Series Expansion for FO Element Realization

The generating functions are essentially a $s \leftrightarrow z$ mapping relation. For band-limited rational approximation of fractional order elements, higher order transfer functions in discrete time need to be found out which approximately mimic the constant phase property of the corresponding FO element, within a chosen frequency range. The higher order approximation is done using various recursive formulae for calculating the position of the poles and zeros of the discrete realization. The order of the realized filter is basically a trade-off between the accuracy in frequency domain (obtained by increased number of poles and zeros) and the ease of hardware implementation (obtained by lower number of poles and zeros). Now, the various expansion techniques are listed below.

- *Power Series Expansion* (*PSE*) Let us now, consider a fractional order differ-integrator

$$G(s) = s^{\pm\gamma}, \gamma \in [0, 1] \subseteq \mathbb{R}_+ \tag{2.26}$$

Using the simple Euler's discretization formula the FO differ-integrator can be approximated via the PSE as

$$T^{\mp\gamma} PSE \left\{ \left(1 - z^{-1}\right)^{\pm\gamma} \right\}$$ (2.27)

Performing PSE on the FO differ-integrator gives (Yang Quan et al. 2009)

$$\nabla_T^{\pm\gamma} f(nT) = T^{\pm\gamma} \sum_{k=0}^{\infty} (-1)^k \binom{\mp\gamma}{k} f((n-k)T)$$ (2.28)

It is evident that the PSE yields polynomial functions or finite impulse response (FIR) type digital filters.

- *Continued Fraction Expansion (CFE)* In most cases rational approximations which yield both poles and zeros are better than PSE based realization which yields FIR filter structure. These rational approximations converge rapidly and have a wider domain of convergence in the complex plane. A smaller set of coefficients are also required for obtaining a good approximation in case of CFE. The CFE of fractional power of a generating function can be expressed as a rational transfer function

$$G(z) \simeq a_0(z) + \cfrac{b_1(z)}{a_1(z) + \cfrac{b_2(z)}{a_2(z) + \cfrac{b_2(z)}{a_3(z) + \cdots}}}$$

$$= a_0(z) + \frac{b_1(z)}{a_1(z)} + \frac{b_2(z)}{a_2(z)} + \frac{b_3(z)}{a_3(z)} + \cdots$$ (2.29)

where a_i, b_i are either rational functions of the variable or constants.

- *MacLaurin Series Expansion* The MacLaurin series expansion of $(x + a)^v$ is given by

$$\sum_{i=0}^{\infty} a^{v-i} \frac{\Gamma(v+1)}{\Gamma(i+1)\Gamma(v-i+1)} x^i$$ (2.30)

For example (Valerio and da Costa 2005),

$$\left(1 - z^{-1}\right)^v = \sum_{i=0}^{\infty} (-1)^i \frac{\Gamma(v+1)}{\Gamma(i+1)\Gamma(v-i+1)} z^{-i}$$ (2.31)

$$\left(1 + z^{-1}\right)^{-v} = \sum_{i=0}^{\infty} \frac{\Gamma(-v+1)}{\Gamma(i+1)\Gamma(-v-i+1)} z^{-i}$$ (2.32)

Other expansions like Taylor Series can be used with higher order generating function to get a better approximation (Gupta et al. 2011; Visweswaran et al. 2011).

2.3 Discrete Time Fractional Order Models

2.3.1 ARFIMA Models

Conventional discrete time system models are mostly moving average (Finite Impulse Response or all zero) and AutoRegressive (Infinite Impulse Response or all pole) type. Generalization of these models is the Auto Regressive Moving Average (ARMA) which is a pole-zero (IIR) transfer function.

The Moving Average (MA) model with q terms $MA\,(q)$ is of the form:

$$y_t = x_t + \theta_1 x_{t-1} + \cdots + \theta_q x_{t-q} \tag{2.33}$$

For the coefficients $\theta_i, i \in \{1, \ldots, q\} \subseteq \mathbb{Z}_+$. The Auto Regressive (AR) models with p terms $AR\,(p)$ is of the form:

$$y_t = \phi_1 y_{t-1} + \cdots + \phi_p y_{t-p} \tag{2.34}$$

for some coefficients $\phi_j, j \in \{1, \ldots, p\} \subseteq \mathbb{Z}_+$.

The AutoRegressive Moving Average (ARMA) model with p AutoRegressive terms and q Moving Average terms $ARMA\,(p, q)$ is given by:

$$y_t = \left(\phi_1 y_{t-1} + \cdots + \phi_p y_{t-p}\right) + \left(x_t + \theta_1 x_{t-1} + \cdots + \theta_q x_{t-q}\right) \tag{2.35}$$

For a time series X_t and zero mean white Gaussian noise ε_t the ARIMA (p, q) model can be represented as:

$$\left(1 - \sum_{i=1}^{p} \phi_i L^i\right) X_t = \left(1 + \sum_{i=1}^{q} \theta_i L^i\right) \varepsilon_t \tag{2.36}$$

where L is the lag operator satisfying the expression $LX_t = X_{t-1}$. The AutoRegressive Integrated Moving Average (ARIMA) model $ARIMA\,(p, d, q)$ is a generalization of the $ARMA\,(p, q)$ model with a differencing parameter $d \in \mathbb{Z}_+$. The Eq. 2.36 for the ARIMA model can be extended to the ARFIMA model to obtain:

$$\left(1 - \sum_{i=1}^{p} \phi_i L^i\right)(1 - L)^d X_t = \left(1 + \sum_{i=1}^{q} \theta_i L^i\right) \varepsilon_t, \; (1 - L)^d = \sum_{k=0}^{\infty} \binom{d}{k} (-L)^k \tag{2.37}$$

The Auto Regressive Fractional Integrated Moving Average (ARFIMA) model is an extension of the ARIMA model where the degree of differencing d can have fractional values (Hosking 1981). ARFIMA models have found applications in economics, biology, remote sensing, network traffic modeling and similar areas (Sheng and Chen 2011).

An $ARFIMA\,(p, d, q)$ has the following properties (Chen et al. 2007) depending on various values of d.

1. For $d = 1/2$, the *ARFIMA* $(p, -1/2, q)$ process is stationary but not invertible.
2. For $-1/2 < d < 0$, the *ARFIMA* (p, d, q) process has short memory and decay monotonically and hyperbolically to zero.
3. For $d = 0$, the *ARFIMA* $(p, 0, q)$ process can be white noise.
4. For $0 < d < 1/2$, the *ARFIMA* (p, d, q) process is a stationary process with long memory. This model is useful in modeling long-range dependence (LRD). The autocorrelation of a LRD time-series decays slowly as a power law function.
5. For $d = 1/2$, the *ARFIMA* $(p, 1/2, q)$ process is a discrete time 1/f noise.

2.3.2 Discrete Time FO Filters

Digital realization of FO filters in FIR and IIR structure is an emerging area of research. Some recent developments include fractional second order filter (Li et al. 2011a), variable order operator (Chan et al. 2010a; Valério and Sá da Costa 2011; Lorenzo and Hartley 2002; Tseng 2006, 2008; Chan et al. 2010b; Charef and Bensouici 2011), distributed order operator (Li et al. 2011b), variable order fractional delay filter (Shyu et al. 2009a, b; Soo-Chang et al. 2010; Tseng 2007b), two-dimensional FO digital differentiator (Chang 2009) and improvement of FO digital differ-integrators (Chien-Cheng 2001, 2006; Tseng 2007a).

2.4 Fractional Order Convolution and Correlation

R_ρ^ϕ is the fractional shift operator and is a generalization of the unitary time shift operator $(T_\tau s)(t) = s(t - \tau)$, R_ρ^ϕ describes a signal shifted along the arbitrary orientation ϕ of the time–frequency plane by a radial distance ρ. R_ρ^ϕ operating on the time domain signal $s(t)$ is given by

$$\left(R_\rho^\phi s\right)(t) = s\left(t - \rho \cos \phi\right) e^{-j2\pi \left(\rho^2/2\right) \cos \phi \sin \phi + j2\pi\rho \sin \phi} \qquad (2.38)$$

The additivity and inversion properties for R_ρ^ϕ can be stated as

$$R_{\rho 1}^\phi R_{\rho 2}^\phi = R_{\rho 1 + \rho 2}^\phi \qquad (2.39)$$

$$\left(R_\rho^\phi\right)^{-1} = R_{-\rho}^\phi \qquad (2.40)$$

$\mathcal{F}_\phi$ is the Fractional Fourier Transform (FrFT) operator associated with angle ϕ. For $\phi = \pi/2$.

It results in the conventional Fourier Transform, and for $\phi = \pi$.

$\mathcal{F}_\pi$ is the axis-reversal operator $(\mathcal{F}_\pi s)(t) = s(-t)$.

It is actually the application of the conventional Fourier Transform twice on the signal. For further details on the FrFT please see Chap. 4 of this book.

The fractional convolution associated with the angle ϕ is found by computing the inner product of the input signal $s(t)$ with the axis-reversed, complex-conjugated, and fractionally shifted version of the function $h(t)$ as

$$
\begin{aligned}
\left(s \overset{*}{\phi} h\right)(r) &= \left\langle s, R_r^\phi \mathcal{F}_\pi h^* \right\rangle = \left\langle s, R_r^\phi \tilde{h} \right\rangle \\
&= e^{j2\pi \frac{r^2}{2} \cos\phi \sin\phi} \int s(\beta) h(r\cos\phi - \beta) e^{-j2\pi\beta r \sin\phi} d\beta
\end{aligned}
\tag{2.41}
$$

where $\tilde{h}(t) = h^*(-t)$. This reduces to the integer order convolution for $\phi = 0$

$$
\left(s \overset{*}{0} h\right)(t) = \int s(\beta) h(t - \beta) d\beta
\tag{2.42}
$$

The fractional order cross-correlation between $s(t)$ and $h(t)$ can be defined by replacing T_τ with R_ρ^ϕ in the definition of the LTI cross-correlation equation

$$
\left(s \overset{*}{0} h\right)(\tau) = \langle s, T_\tau h \rangle = \int s(\beta) h^*(\beta - \tau) d\beta
\tag{2.43}
$$

Thus the fractional order cross-correlation can be defined as

$$
\left(s \overset{*}{\phi} h\right)(\rho) = e^{j2\pi \frac{\rho^2}{2} \cos\phi \sin\phi} \int s(\beta) h^*(\beta - \rho\cos\phi) e^{-j2\pi\beta\rho \sin\phi} d\beta
\tag{2.44}
$$

which reduces to the integer order definition of cross-correlation in (2.43) for $\phi = 0$.

The fractional order autocorrelation is defined as

$$
\left(s \overset{*}{\phi} s\right)(\rho) = e^{j2\pi \frac{\rho^2}{2} \cos\phi \sin\phi} \int s(\beta) s^*(\beta - \rho\cos\phi) e^{-j2\pi\beta\rho \sin\phi} d\beta
\tag{2.45}
$$

which reduces to the integer order definition of autocorrelation for $\phi = 0$

$$
\left(s \overset{*}{0} s\right)(\tau) = \int s(\beta) s^*(\beta - \tau) d\beta
\tag{2.46}
$$

The defining property of an LTI system and their associated LTI operations is their covariance to time shifts. The LTI convolution fully reflects the effect of the unitary time shift operator T_k on the signal $s(t)$ as follows,

$$
\begin{aligned}
\left(T_k s \overset{*}{0} h\right)(t) &= \left\langle T_k s, T_t \tilde{h} \right\rangle = \left\langle s, T_{-k} T_t \tilde{h} \right\rangle \\
&= \left\langle s, T_{t-k} \tilde{h} \right\rangle = \left(s \overset{*}{0} h\right)(t - k)
\end{aligned}
\tag{2.47}
$$

The same covariance property is also valid for the LTI cross-correlation and the LTI autocorrelation.

Similar to time-shift covariance of LTI systems, fractional convolution and correlation operations are covariant to fractional shifts represented by the unitary fractional-shift operator R_ρ^ϕ.

For fractional convolution this can be shown by

$$\left(R_k^\phi s_\phi^* h\right)(r) = \left\langle R_k^\phi s, R_r^\phi \widetilde{h}\right\rangle = \left\langle s, R_{-k}^\phi s, R_r^\phi \widetilde{h}\right\rangle$$
$$= \left\langle s, R_{r-k}^\phi \widetilde{h}\right\rangle = \left(s_\phi^* h\right)(r-k) \qquad (2.48)$$

where we used the fact that R_ρ^ϕ is unitary and $\left(R_r^\phi\right)^{-1} = R_{-r}^\phi$ as given by the inversion property in (2.40).

Fractional correlation operators are also covariant to fractional shifts in the same fashion, i.e.

$$\left(R_k^\phi s_\phi^* h\right)(\rho) = \left(s_\phi^* h\right)(\rho - k) \qquad (2.49)$$

For further details please see Akay and Boudreaux-Bartels (2001) and Torres et al. (2010).

2.5 Calculation of H₂ and H∞ Norms of Fractional Order Systems

For FO LTI systems, the H_2 and H_∞ norms are similar to their corresponding integer order analogues with slight modifications as detailed in Malti et al. (2003) and Sabatier et al. (2005).

2.5.1 H₂ Norm of FO Systems

The H_2 norm of a transfer function $F(s)$, i.e., $\|F\|_{H_2} := \|F\|_2$ which is analogous to the impulse response energy $\|f\|_2$ or L_2 norm of a real signal $f(t)$ can be stated as:

$$\|f(t)\|_2 = \left(\int_0^\infty f^T(t) f(t) \, dt\right)^{1/2}$$
$$= \left(\frac{1}{2\pi} \int_{-\infty}^\infty F^H(j\omega) F(j\omega) \, d\omega\right)^{1/2} = \|F(j\omega)\|_2 \qquad (2.50)$$

If it be assumed that $F(s)$ be a Laplace-domain BIBO-stable explicit commensurate fractional order (n) transfer function and q, ρ are the integer and non-integer parts of $1/n$ respectively, i.e. $q := \left\lfloor \frac{1}{n} \right\rfloor$, $\rho := \frac{1}{n} - \left\lfloor \frac{1}{n} \right\rfloor$ and let,

$$G\left(\omega^n\right) := F\left(j\omega\right), \quad \frac{A\left(x\right)}{B\left(x\right)} := G\left(x\right)\overline{G\left(x\right)} \tag{2.51}$$

Then the H_2 norm of $F\left(s\right)$ is given by the following conditions:

1. If $\deg\left(B\right) \leq \deg\left(A\right) + 1/n$, then $\|F\|_2 = \infty$.
2. If $\deg\left(B\right) > \deg\left(A\right) + 1/n$ and $\rho \neq 0$, then

$$\|F\|_2 = \frac{\sum_{k=1}^{r}\sum_{l=1}^{v_k} (-1)^{l-1} a_{k,l} s_k^{\rho-1} \binom{\rho-1}{l-1}}{n\sin\left(\rho\pi\right)} \tag{2.52}$$

where $a_{k,l}$, $(-s_k)$ and v_k are co-efficients, poles and their multiplicities of the partial fraction expansion of

$$x^q \frac{A\left(x\right)}{B\left(x\right)} = \sum_{k=1}^{r}\sum_{l=1}^{v_k} \frac{a_{k,l}}{\left(x+s_k\right)^l}$$

3. If $\deg\left(B\right) > \deg\left(A\right) + \frac{1}{n}$ and $\rho = 0$:

$$\|F\|_2^2 = \sum_{k=2}^{r} \frac{c_k\left(\ln\left(s_k\right) - \ln\left(s_1\right)\right)}{n\pi\left(s_k - s_1\right)} + \sum_{k=1}^{r}\sum_{l=2}^{v_k} \frac{b_{k,l} s_k^{1-l}}{n\pi\left(l-1\right)} \tag{2.53}$$

where $(-s_k)$ and v_k are poles and their multiplicities of $x^{q-1}Q\left(x\right)$. s_1 is an arbitrary chosen pole. c_k and $b_{k,l}$ represent coefficients of the following expansion of $x^{q-1}Q\left(x\right)$:

$$x^{q-1}\frac{A\left(x\right)}{B\left(x\right)} = \sum_{k=2}^{r} \frac{c_k}{\left(x+s_1\right)\left(x+s_k\right)} + \sum_{k=1}^{r}\sum_{l=2}^{v_k} \frac{b_{k,l}}{\left(x+s_k\right)^l} \tag{2.54}$$

All poles of order one are in the left sum; all poles of order greater than one are in the right double sum.

2.5.2 H_∞ Norm of FO Systems

The H_∞-norm of a continuous LTI system model $G\left(s\right)$ is defined through the L_2 norm as:

$$\|G\left(s\right)\|_\infty = \sup_{U(s)\in H_2} \frac{\|Y\left(s\right)\|_2}{\|U\left(s\right)\|_2} \tag{2.55}$$

Here, $Y\left(s\right)$, $U\left(s\right) \in H_2$ represent the Laplace transform of the output signal and input signal respectively. Also,

$$\|G(s)\|_\infty = \sup_{\omega \in \mathbb{R}} \sigma_{\max}(G(j\omega)) \tag{2.56}$$

where $\sigma_{\max}$ denotes the greatest singular value of $G(j\omega)$ defined by

$$\sigma(G(j\omega)) = \sqrt{spec\left(G^T(-j\omega)\,G(j\omega)\right)} \tag{2.57}$$

The H_∞-norm of a fractional system may not always be bounded. For the commensurate fractional state space system S_f

$$\left(S_f\right) : \begin{cases} D^v x(t) = Ax(t) + Bu(t) \\ y(t) = Cx(t) + Du(t) \end{cases} \tag{2.58}$$

where $v \in \mathbb{R}_+$ denotes the fractional order of the system,

$$A \in \mathbb{R}^{n \times n},\, B \in \mathbb{R}^{p \times l},\, C \in \mathbb{R}^{m \times n} \text{ and } D \in \mathbb{R}^{m \times p}$$

The H_∞-norm is bounded by a real positive number γ if and only if the eigenvalues of the matrix A_γ lie in the stable domain defined by $\{s \in \mathbb{C} : |\arg(s)| > v\pi/2\}$. Here,

$$A_\gamma = \begin{pmatrix} A + B\left(\gamma^2 I - D^T D\right)^{-1} D^T C & e^{-vj\pi} B\left(\gamma^2 I - D^T D\right)^{-1} B^T \\ C^T\left(I + D\left(\gamma^2 - D^T D\right)^{-1} D^T\right) C\, e^{-vj\pi} & \left(A^T + C^T D\left(\gamma^2 I - D^T D\right)^{-1} B^T\right) \end{pmatrix}$$

References

Akay, O., Boudreaux-Bartels, G.F.: Fractional convolution and correlation via operator methods and an application to detection of linear FM signals. IEEE Trans. Signal Process. **49**(5), 979–993 (2001)

Barbosa, R.S., Tenreiro Machado, J.A., Silva, M.F.: Time domain design of fractional differintegrators using least-squares. Signal Process. **86**(10), 2567–2581 (2006). doi:10.1016/j.sigpro.2006.02.005

Bouafoura, M.K., Moussi, O., Braiek, N.B.: A fractional state space realization method with block pulse basis. Signal Process. **91**(3), 492–497 (2011). doi:10.1016/j.sigpro.2010.04.010

Carlson, G., Halijak, C.: Approximation of fractional capacitors (1/s)^(1/n) by a regular newton process. IEEE Trans. Circuit Theory **11**(2), 210–213 (1964)

Chan, C.-H., Shyu, J.-J., Hsin-Hsyong Yang, R.: Iterative design of variable fractional-order IIR differintegrators. Signal Process. **90**(2), 670–678 (2010). doi:10.1016/j.sigpro.2009.08.006

Chan, C.-H., Shyu, J.-J., Yang, R.H.-H.: A new structure for the design of wideband variable fractional-order FIR differentiators. Signal Process. **90**(8), 2594–2604 (2010). doi:10.1016/j.sigpro.2010.03.005

Chang, W.-D.: Two-dimensional fractional-order digital differentiator design by using differential evolution algorithm. Digit. Signal Process. **19**(4), 660–667 (2009). doi:10.1016/j.dsp.2008.12.004

Charef, A., Bensouici, T.: Design of digital FIR variable fractional order integrator and differentiator. Signal Image Video Process. 1–11 (2011, in press). doi:10.1007/s11760-010-0197-1

Charef, A., Sun, H.H., Tsao, Y.Y., Onaral, B.: Fractal system as represented by singularity function. IEEE Trans. Autom. Control **37**(9), 1465–1470 (1992)

Chen, Y., Sun, R., Zhou, A.: An overview of fractional order signal processing (FOSP) techniques. ASME Conf. Proc. **2007**(4806X), 1205–1222 (2007). doi:10.1115/detc2007-34228

Chen, Y., Vinagre, B.M.: A new IIR-type digital fractional order differentiator. Signal Process. **83**(11), 2359–2365 (2003). doi:10.1016/s0165-1684(03)00188-9

Chen, Y., Vinagre, B.M., Podlubny, I.: Continued fraction expansion approaches to discretizing fractional order derivatives—an expository review. Nonlinear Dyn. **38**(1), 155–170 (2004). doi:10.1007/s11071-004-3752-x

Chien-Cheng, T.: Design of fractional order digital FIR differentiators. IEEE Signal Process. Lett. **8**(3), 77–79 (2001)

Chien-Cheng, T.: Improved design of digital fractional-order differentiators using fractional sample delay. IEEE Trans. Circuits Syst. I Regul. Papers **3**(1), 193–203 (2006)

Elwakil, A.S.: Fractional-order circuits and systems: an emerging interdisciplinary research area. IEEE Circuits Syst. Mag. **10**(4), 40–50 (2010)

Gupta, M., Varshney, P., Visweswaran, G.S.: Digital fractional-order differentiator and integrator models based on first-order and higher order operators. Int. J. Circuit Theory Appl. **39**(5), 461–474 (2011). doi:10.1002/cta.650

Hosking, J.R.M.: Fractional differencing. Biometrika **68**(1), 165–176 (1981)

Krishna, B.T.: Studies on fractional order differentiators and integrators: a survey. Signal Process. **91**(3), 386–426 (2011). doi:10.1016/j.sigpro.2010.06.022

Li, Y., Sheng, H., Chen, Y.: Analytical impulse response of a fractional second order filter and its impulse response invariant discretization. Signal Process. **91**(3), 498–507 (2011). doi:10.1016/j.sigpro.2010.01.017

Li, Y., Sheng, H., Chen, Y.Q.: On distributed order integrator/differentiator. Signal Process. **91**(5), 1079–1084 (2011). doi:10.1016/j.sigpro.2010.10.005

Lorenzo, C.F., Hartley, T.T.: Variable order and distributed order fractional operators. Nonlinear Dyn. **29**(1), 57–98 (2002). doi:10.1023/a:1016586905654

Magin, R., Ortigueira, M.D., Podlubny, I., Trujillo, J.: On the fractional signals and systems. Signal Process. **91**(3), 350–371 (2011). doi:10.1016/j.sigpro.2010.08.003

Maione, G.: Continued fractions approximation of the impulse response of fractional-order dynamic systems. IET Control Theory Appl. **2**(7), 564–572 (2008)

Malti, R., Aoun, M., Cois, O., Oustaloup, A., Levron, F.: H[sub 2] norm of fractional differential systems. ASME Conf. Proc. **2003**(37033), 729–735 (2003). doi:10.1115/detc2003/vib-48387

Matignon, D.: Stability properties for generalized fractional differential systems. ESAIM Proc. **5**, 145–158 (1998)

Maundy, B., Elwakil, A.S., Freeborn, T.J.: On the practical realization of higher-order filters with fractional stepping. Signal Process. **91**(3), 484–491 (2011). doi:10.1016/j.sigpro.2010.06.018

Merrikh-Bayat, F., Afshar, M., Karimi-Ghartemani, M.: Extension of the root-locus method to a certain class of fractional-order systems. ISA Trans. **48**(1), 48–53 (2009). doi:10.1016/j.isatra.2008.08.001

Monje, C.A., Chen, Y.Q., Vinagre, B.M., Xue, D., Feliu, V. Fractional-Order Systems and Controls: Fundamentals and Applications. Springer, Heidelberg (2010)

Ortigueira, M.D.: Introduction to fractional linear systems. Part 1. Continuous-time case. IEE Proc. Vis. Image Signal Process. **147**(1), 62–70 (2000)

Oriquetra, M.D.: Introduction to fractional linear systems. Part 2. Discrete-time case. IEE Proc. Vis. Image Signal Process. **1**(147), 71–78 (2000)

Ortigueira, M.D.: On the initial conditions in continuous-time fractional linear systems. Signal Process. **83**(11), 2301–2309 (2003). doi:10.1016/s0165-1684(03)00183-x

Ortigueira, M.D.: A coherent approach to non-integer order derivatives. Signal Process. **86**(10), 2505–2515 (2006). doi:10.1016/j.sigpro.2006.02.002

Ortigueira, M.D.: An introduction to the fractional continuous-time linear systems: the 21st century systems. IEEE Circuits Syst. Mag. **8**(3), 19–26 (2008)

Ortigueira, M.D.: On the fractional linear scale invariant systems. IEEE Trans. Signal Process. **58**(12), 6406–6410 (2010)

Oustaloup, A., Levron, F., Mathieu, B., Nanot, F.M.: Frequency-band complex noninteger differentiator: characterization and synthesis. IEEE Trans. Circuits Syst. I Fundam. Theory Appl. **47**(1), 25–39 (2000)

Rachid, M., Maamar, B., Said, D.: Comparison between two approximation methods of state space fractional systems. Signal Process. **91**(3), 461–469 (2011). doi:10.1016/j.sigpro.2010.03.006

Sabatier, J., Moze, M., Oustaloup, A.: On fractional systems H< inf> ∞< /inf>,-Norm computation. In: 44th IEEE conference on decision and control, 2005 and 2005 european control conference. CDC-ECC '05. 12–15 December 2005, pp. 5758–5763 (2005)

Sheng, H., Chen, Y.: FARIMA with stable innovations model of Great Salt Lake elevation time series. Signal Process. **91**(3), 553–561 (2011). doi:10.1016/j.sigpro.2010.01.023

Shyu, J.-J., Pei, S.-C., Chan, C.-H.: An iterative method for the design of variable fractional-order FIR differintegrators. Signal Process. **89**(3), 320–327 (2009a). doi:10.1016/j.sigpro.2008.09.009

Shyu, J.-J., Pei, S.-C., Chan, C.-H.: Minimax phase error design of allpass variable fractional-delay digital filters by iterative weighted least-squares method. Signal Process. **89**(9), 1774–1781 (2009b). doi:10.1016/j.sigpro.2009.03.021

Siami, M., Saleh Tavazoei, M., Haeri, M.: Stability preservation analysis in direct discretization of fractional order transfer functions. Signal Process. **91**(3), 508–512 (2011). doi:10.1016/j.sigpro.2010.06.009

Soo-Chang, P., Peng-Hua, W., Chia-Huei, L.: Design of fractional delay filter differintegrator fractional hilbert transformer and differentiator in time domain with peano kernel. IEEE Trans. Circuits Syst. I Regul. Papers **57**(2), 391–404 (2010)

Torres, R., Pellat-Finet, P., Torres, Y.: Fractional convolution, fractional correlation and their translation invariance properties. Signal Process. **90**(6), 1976–1984 (2010). doi:10.1016/j.sigpro.2009.12.016

Trigeassou, J.C., Maamri, N.: Initial conditions and initialization of linear fractional differential equations. Signal Process. **91**(3), 427–436 (2011). doi:10.1016/j.sigpro.2010.03.010

Trigeassou, J.C., Maamri, N., Sabatier, J., Oustaloup, A.: A Lyapunov approach to the stability of fractional differential equations. Signal Process. **91**(3), 437–445 (2011). doi:10.1016/j.sigpro.2010.04.024

Tseng, C.-C.: Design of variable and adaptive fractional order FIR differentiators. Signal Process. **86**(10), 2554–2566 (2006). doi:10.1016/j.sigpro.2006.02.004

Tseng, C.-C.: Design of FIR and IIR fractional order Simpson digital integrators. Signal Process. **87**(5), 1045–1057 (2007). doi:10.1016/j.sigpro.2006.09.006

Tseng, C.-C.: Designs of fractional delay filter, Nyquist filter, lowpass filter and diamond-shaped filter. Signal Process. **87**(4), 584–601 (2007). doi:10.1016/j.sigpro.2006.06.016

Tseng, C.-C.: Series expansion design of variable fractional order integrator and differentiator using logarithm. Signal Process. **88**(9), 2278–2292 (2008). doi:10.1016/j.sigpro.2008.03.012

Valerio, D., da Costa, J.S.: Time-domai implementatio of fractional order controllers. IEE Proc. Control Theory Appl. **152**(5), 539–552 (2005)

Valério, D., Sáda Costa, J.: Variable-order fractional derivatives and their numerical approximations. Signal Process. **91**(3), 470–483 (2011). doi:10.1016/j.sigpro.2010.04.006

Visweswaran, G.S., Varshney, P., Gupta, M.: New approach to realize fractional power in domain at low frequency. IEEE Trans. Circuits Syst. II Express Br. **58**(3), 179–183 (2011)

Yang Quan, C., Petras, I., Dingyu, X.: Fractional order control—A tutorial. In: American control conference, 2009. ACC '09. 10–12 June 2009, pp. 1397–1411 (2009)

Chapter 3
Long Range Dependence, Stable Distributions and Self-Similarity

Abstract This chapter looks at Long Range Dependent models which are useful in modeling real world signals having outliers or long tailed statistical distributions. The concept of stable distribution is introduced and many standard distributions like symmetric αstable, Gaussian, Cauchy, etc. are obtained from the generalized description of the family of stable distributions. Self-similarity for random processes is next introduced and the ideas of fractional Brownian motion, fractional and multi-fractional Gaussian noise, etc. are discussed in this context. The Hurst parameter which is a measure of self-similarity is discussed next and a few methods to estimate it are subsequently outlined.

Keywords α-Stable distributions · Fractional Brownian motion (fBm) · Fractional Gaussian noise (fGn) · Self similarity · Fractal dimension · Hurst parameter

3.1 Introduction

In statistical signal processing, models of self similar processes are stochastic and a scaling in time is equivalent to a suitable scaling in space. The link between the two types of scaling is determined by a constant called Hurst exponent. Depending on the value of the Hurst parameter the process can show Long Range Dependence (LRD) or long memory. Memory here implies something that lasts and the notion of memory is related to the connection between certain observations and those occurring after the passage of a certain interval of time (i.e. the shifted process). LRD is linked to the fact that the rate of decay of the statistical dependence of the process is more slower like a power-law rather than an exponential one.

LRD is found in atmospheric temperature and ozone fluctuations as reported in Varotsos and Kirk-Davidoff (2006). It can also be found in DNA sequences (Karmeshu and Krishnamachari 2004), hydrology (Montanari and Toth 2007), network traffic modeling (Grossglauser and Bolot 1999), econometrics, stock market

S. Das and I. Pan, *Fractional Order Signal Processing*,
SpringerBriefs in Applied Sciences and Technology,
DOI: 10.1007/978-3-642-23117-9_3, © The Author(s) 2012

fluctuations and many more (Doukhan et al. 2003). Thus this type of model can be applied to a wide variety of processes for their simulation and analysis.

3.2 Rational Transfer Function Models of Signals and its Extension to ARFIMA Models

A time series is essentially an ordered sequence of observations where the successive elements are related to each other. Past samples may be used to predict the future samples in the series if they are dependent. However, for stochastic time series this prediction is never exact and the degree of predictability is related to the dependence between the successive samples. The white noise is a special case where each sample is independent of the others and the signal is completely unpredictable.

A model is a mathematical representation of a signal which captures the important characteristics of the signal. Assuming that a given sample $x(n)$, at time n can be approximated by the weighted value (a) of the previous sample, we have $x(n) \approx ax(n-1)$, where a is assumed to be a constant. An excitation term $w(n)$ can be added to make the above relationship exact, resulting in

$$x(n) = ax(n-1) + w(n) \tag{3.1}$$

This can be represented in a discrete time transfer function form as

$$H(z) = \frac{X(z)}{W(z)} = \frac{1}{1 - az^{-1}} \tag{3.2}$$

Using inverse transformation and series expansion we obtain

$$x(n) = w(n) + aw(n-1) + a^2 w(n-2) + \cdots \tag{3.3}$$

which indicates that the model generates a time series with exponentially decaying dependence.

Any rational transfer function model has a dependence structure or memory that decays exponentially with time and is known as AutoRegressive Moving-Average (ARMA) model or pole zero model. The impulse response and the autocorrelation sequence of a pole-zero model decay exponentially with time, and can be expressed as

$$|h(n)| \le C_h \zeta^{-n}, \quad |\rho(l)| \le C_\rho \zeta^{-l} \tag{3.4}$$

where $C_h, C_\rho > 0$ and $0 < \zeta < 1$

Hence these types of pole-zero models may be used to model stationary signals with short memory.

If one of the poles moves quite close to the unit circle, a longer impulse response or autocorrelation can be obtained. Hence models with unit poles can be used to represent some kinds of non-stationary processes with trends. However, even these models cannot effectively capture the characteristics of a time series whose autocorrelation decays asymptotically as a power law.

For a unit pole ($a = 1$), a perpetual constant dependence is obtained but the model is unstable as the output is unbounded. However, a stable model with long memory can be obtained by creating a fractional unit pole (Hosking 1981), that is by including a fractional power in (3.2).

$$H_d(z) = \frac{1}{(1 - az^{-1})^d} = 1 + dz^{-1} + \frac{d(d+1)}{2!}z^{-2} + \cdots, \quad -\frac{1}{2} < d < \frac{1}{2} \quad (3.5)$$

Thus,

$$x(n) = w(n) + dw(n-1) + \frac{d(d+1)}{2!}w(n-2) + \cdots \quad (3.6)$$

The weights in (3.6) decay according to n^{d-1} as $n \to \infty$ i.e. the dependence decays asymptotically as a power law or hyperbolically. Although the model in (3.6) is characterized by a single parameter d, it is actually an infinite-order convolution summation and hence its practical realization requires a rational model approximation which can be implemented as a difference equation. The process generated by (3.6) is known as fractionally differenced Gaussian noise when $w(n)$ is a sequence of independent Gaussian random variables.

In time series analysis, rational model which have one or more fractional poles is known as Fractional AutoRegressive Integrated Moving Average (FARIMA) models. Their importance in modeling time series stems from the fact that their short-term dependence is exponential while their long-term dependence is hyperbolic. When these fractional pole-zero models are excited by white noise, random signals are generated which can be effectively modeled as AutoRegressive Fractionally Integrated Moving Average (ARFIMA) process and have strong correlation even if the samples are far apart. These FARIMA or ARFIMA models exhibit long term persistence and cannot be effectively modeled using the conventional rational pole zero models.

LRD can be modeled in continuous time by using a fractional pole i.e. $1/s^\beta$, $\beta > 0$. This is actually a fractional integrator for $0 < \beta < 1$ and an integrator for $\beta = 1$. Since the inverse Laplace transform of the fractional integrator is given by

$$\mathcal{L}^{-1}\left\{\frac{1}{s^\beta}\right\} = t^{\beta-1} \quad (3.7)$$

the memory of a continuous-time system with impulse response $h_\beta = t^{\beta-1}$ for $t \geq 0$ and $h_\beta = 0$ for $t < 0$, decays hyperbolically. The output of this kind of system to white Gaussian noise results in a non-stationary process called fractional Brownian motion (fBm).

The stationary discrete-time process, obtained by sampling the fractional Brownian motion process at equal intervals and computing the one-step increments, is known as fractional Gaussian noise (fGn). Both fBm and fGn have long memory and have wide applicability in signal processing domain.

3.3 Few Characteristics of Fractional Pole Zero Models

Some important properties of these models like impulse response, power spectrum, correlation, etc. are discussed next (Manolakis et al. 2005). The impulse response of the fractional pole model in (3.5) is given by

$$h_d(n) = \frac{\Gamma(n+d)}{\Gamma(n+1)\Gamma(d)} \tag{3.8}$$

for $n \geq 0$ and $h_d(n) = 0$ for $n < 0$. As $n \to \infty$ the impulse response

$$h(n) \sim \frac{1}{(d-1)!} n^{d-1} \tag{3.9}$$

Thus the system is not BIBO stable for $d > 0$. However, the fractional pole model is termed as minimum-phase for $-1/2 < d < 1/2$ even though $h_d(n)$ does not converge absolutely. In this range the model exhibits mean square convergence.

The complex power spectrum of the model is $R_x(z) = \sigma_w^2 R_h(z)$ where

$$R_h(z) = H(z)H(z^{-1}) = \frac{1}{(1-z^{-1})^d(1-z)^d} \tag{3.10}$$

Putting $z = e^{j\omega}$ the power spectrum is obtained as

$$R_h(e^{j\omega}) = \frac{1}{[2\sin(\omega/2)]^{2d}} \qquad -\pi < \omega < \pi \tag{3.11}$$

$R_h(e^{j\omega})$ is finite only for $d \leq 0$. Also as $\omega \to 0$, $R_h(e^{j\omega}) \sim \frac{1}{\omega^{2d}}$.

The normalized autocorrelation is given by

$$\rho_h(l) = \frac{(d+l-1)!}{(d-1)!(l-d)!} \tag{3.12}$$

In (3.12) as $l \to \infty$, $\rho_h(l) \sim C_d l^{2d-1}$. For $0 < d < 1/2$ the autocorrelation sequence decay monotonically to zero. In this case $\sum_{l=-\infty}^{\infty} |\rho(l)| = \infty$ and $R(e^{j\omega}) \to \infty$ as $\omega \to \infty$. The spectrum is dominated by low-frequency components (low-pass), and the divergence at $\omega = 0$ causes the long memory behavior. The system acts as a fractional integrator. For $-1/2 < d < 0$ the autocorrelation sequence decays monotonically to zero. In this case, $\sum_{l=-\infty}^{\infty} |\rho(l)| < \infty$, $R(e^{j0}) = \sum_{l=-\infty}^{\infty} |\rho(l)| = 0$, and the spectrum is dominated by high-frequency components (high-pass). The model exhibits short-memory behavior. The system acts as a fractional differentiator.

3.4 Stable Distributions

A random variable is said to have a stable distribution if a linear combination of two independent copies of the variable preserve the shape of the original distribution (up to scale and shift).

A random variable $x(\zeta)$ is stable or stable in the broad sense if, for $x_1(\zeta)$ and $x_2(\zeta)$ independent copies of $x(\zeta)$ and any positive constants a and b, (3.13) holds for some positive c and some $d \in \mathbb{R}$.

$$ax_1(\zeta) + bx_2(\zeta) \overset{d}{=} cx(\zeta) + d \tag{3.13}$$

where, the symbol $\overset{d}{=}$ implies equality in distribution (i.e. both expressions have the same probability law). The random variable is strictly stable or stable in the narrow sense if (3.13) holds with $d = 0$ for all choices of a and b. A random variable is symmetric stable if it is stable and symmetrically distributed around zero, e.g. $x(\zeta) \overset{d}{=} -x(\zeta)$.

There is no closed form expression for probability density function of these stable random variables and are specified by their characteristic function given by

$$\Phi(\xi) = \begin{cases} e^{j\mu\xi - |\sigma\xi|^{\alpha} \cdot \left(1 - j\beta sign(\xi) \tan\left(\frac{\pi\alpha}{2}\right)\right)} & \text{for } \alpha \neq 1 \\ e^{j\mu\xi - |\sigma\xi|^{\alpha} \cdot \left(1 - j\beta\left(\frac{2}{\pi}\right) sign(\xi) \ln\left(|\xi|\right)\right)} & \text{for } \alpha = 1 \end{cases} \tag{3.14}$$

where, $sign(\xi) = \xi/|\xi|$ if $\xi \neq 0$ and zero otherwise.

The parameters in (3.14) can be interpreted as follows:

1. The characteristic exponent α, $0 < \alpha \leq 2$, reflects the shape of the distribution and therefore the flatness of the tails. α is known as the index of stability. A stable random variable $x(\zeta)$ with index α is called α-stable.
2. The symmetry of the distribution is determined by the skewness parameter β, $-1 < \beta < 1$. For $\beta = 0$ the distribution is symmetric, for $\beta < 0$ it is left skewed and for $\beta > 0$ it is right skewed.
3. The range of dispersion of the stable distribution is dictated by the scale parameter σ, $0 \leq \sigma < \infty$.
4. The center of the distribution is determined by the location parameter μ, $-\infty < \mu < \infty$.

For $0 < \alpha < 2$, the tails of a stable distribution decay as a power law, i.e.

$$\Pr\left[|x(\zeta) - \mu| \geq x\right] \simeq \frac{C}{x^{\alpha}} \text{ as } x \to \infty \tag{3.15}$$

where, C is a constant that depends on the scale parameter σ. Due to this nature, α-stable random variables have infinite second-order moments.

A symmetric α-stable distribution is denoted as $S\alpha S$, and its characteristic function is given by

$$\Phi(\xi) = e^{j\mu\xi - |\sigma\xi|^{\alpha}} \tag{3.16}$$

If $x(\zeta)$ is $S\alpha S$ with $\alpha = 2$ in (3.14), a Gaussian distribution is obtained with variance equal to $2\sigma^2$, that is, $N(\mu, 2\sigma^2)$, whose tails decay exponentially and not as a power law. Thus, the Gaussian is the only stable distribution with finite variance. The other stable distributions have infinite variance which implies that the corresponding random variables show large fluctuations. These are helpful in modeling signals which vary between a big range.

If $x(\zeta)$ is $S\alpha S$ with $\alpha = 1$, we have a Cauchy distribution with density

$$f_x(x) = \frac{\sigma/\pi}{(x - \mu)^2 + \sigma^2} \tag{3.17}$$

A standard ($\mu = 0$, $\sigma = 1$) Cauchy random variable $x(\zeta)$ can be generated from a $[0, 1]$ uniform random variable $u(\zeta)$, by using the transformation

$$x = \tan\left[\pi\left(u - \frac{1}{2}\right)\right] \tag{3.18}$$

If $x(\zeta)$ is $S\alpha S$ with $\alpha = 1/2$, we have a Levy distribution, which has both infinite variance and infinite mean. The Probability Density Functions (PDF) of this distribution does not have a functional form and hence must be computed numerically. For further details of stable distributions please refer to Nolan (2003).

3.5 Statistical Self-Similarity of Stochastic Processes

A stochastic process is statistically self-similar or scale-invariant if its statistical properties are invariant under suitable scaling of time and space. Geometric objects (like the Koch curve etc.) which exhibit self similarity are known as fractals (Mandelbrot 1983), whereas random signals which exhibit self similarity are known as random fractals. Small fluctuations at small scales manifest themselves as larger fluctuations at larger scales in a statistical self similar process.

A stochastic process $\{x(t), t \geq 0\}$ is said to be statistically self-similar if for any $c > 0$ there exists $b > 0$ such that

$$x(ct) \stackrel{d}{=} bx(t) \tag{3.19}$$

where, $\stackrel{d}{=}$ denotes equality in all finite dimensional distribution.

Moreover, there exists a unique exponent $H \geq 0$ (known as the Hurst parameter) such that b can be expressed as $b = c^H$. Thus for such a process $x(t)$ a change in time scale can be interpreted as being statistically equivalent to a change in amplitude scale, i.e. the transformation $x(t) \rightarrow c^{-H}x(ct)$ is statistics invariant. Apart from $H = 0$ all self-similar processes are non stationary.

A second order stochastic process $\{x(t), t \geq 0\}$ is wide sense H-Self-similar if the following two conditions hold:

$$E\{x(ct)\} = c^H E\{x(t)\} \quad \forall t > 0, \quad c > 0 \tag{3.20}$$

and

$$E\{x(ct)x(cs)\} = c^{2H} E\{x(t)x(s)\} \quad \forall t > 0, \; s > 0, \; c > 0 \tag{3.21}$$

For strict sense H-self-similarity,

$$x(ct) \overset{d}{=} c^H x(t) \quad \forall c \tag{3.22}$$

All finite-dimensional distributions are equal for strict-sense self-similar processes, while only second-order moments are equal for wide-sense self-similar processes. These moments are as follows

$$\mu_x(t) \overset{\Delta}{=} E\{x(t)\} = c^{-H}\mu_x(ct) \tag{3.23}$$

$$r_x(t_1, t_2) \overset{\Delta}{=} E\{x(t_1)x(t_2)\} = c^{-2H}r_x(ct_1, ct_2) \tag{3.24}$$

The wide sense definition is more general although it excludes self similar processes with infinite second moments (e.g. non-Gaussian stable processes).

3.5.1 Self-Similar Processes with Stationary Increments

A real-valued process $x(t)$ has stationary increments if

$$x(t + \tau) - x(\tau) \overset{d}{=} x(t) - x(0) \quad \forall \tau \tag{3.25}$$

i.e. the distribution of $x(t + \tau) - x(\tau)$ is independent of τ.

A continuous-time stochastic process is self-similar with stationary increments (H-sssi) if and only if it is self-similar with index H and has stationary increments. The mean value, variance and autocorrelation of an H-sssi process are given by the Eqs. 3.26–3.28, respectively

$$\mu_x(t) = 0 \tag{3.26}$$

$$\sigma_x^2(t) = t^{2H}\sigma_H^2 \tag{3.27}$$

$$r_x(t_1, t_2) = \frac{1}{2}\sigma_H^2 \left(|t_1|^{2H} - |t_1 - t_2|^{2H} + |t_2|^{2H}\right) \tag{3.28}$$

where $\sigma_H^2 = E\left\{x^2\,(1)\right\}$.

The power spectrum of self-similar processes follows a power law and is given by

$$R_x\,(F) = \frac{\sigma_H^2}{|F|^{2H+1}} \tag{3.29}$$

where, F is the frequency in cycles per unit of time. Since, $R_x\,(cF) = c^{-(2H+1)} R_x\,(F)$ the process is wide-sense self-similar.

3.5.2 Fractional Brownian Motion

The fractional Brownian motion (fBm), (Mandelbrot and Van Ness 1968) is a special case of the H-sssi process where the probability distribution is Gaussian. It is denoted by $B_H\,(t)$. The fBms are a generalization of the ordinary Brownian motion ($H = 1/2$) with Hurst exponents lying in the range $0 < H < 1$.

For the fBm,

$$E\,[B_H(t)B_H(s)] = \frac{1}{2}\left\{t^{2H} + s^{2H} - |t-s|^{2H}\right\} E\left[B_H(1)^2\right] \tag{3.30}$$

The concept of fBm and fGn can also be described from the viewpoint of fractional calculus (Magin et al. 2011). The fGn, $r(t)$ can be defined as the fractional derivative of the white noise $w(t)$, and is given by,

$$r_\alpha(t) = D^\alpha w(t) \tag{3.31}$$

where D^α is the fractional derivative operator of order α.

For the fGn $r_\alpha(t)$, the fBm $v_\alpha(t), t \geq 0$, is defined as

$$v_\alpha(t) = \int_0^t r_\alpha(\tau)d\tau \tag{3.32}$$

3.5.3 Persistence and Anti-Persistence

The spectrum of the fBm is parabolic for $0 < H < 1/2$ and is known to have anti-persistence property since the increments tend to have opposite sign. This can be obtained by the integration of a stationary fractional noise. A time series data from a stock market fluctuation may exhibit such property. The time domain traces of fBm show more wrinkled nature at lower values of H and are negatively correlated.

The spectrum of the fBm is hyperbolic for $1/2 < H < 1$ and is known to have the persistence property as the increments tend to have the same sign. This can be obtained by the integration of a non-stationary fractional noise. Geographical coastlines exhibit such property. As H increases the time domain traces are smoother and exhibit positive correlation, i.e. they show persistence in the direction in which they are moving.

The ordinary Brownian motion $H = 0.5$ are neutral in persistence, i.e. the time domain traces have zero correlation and do not show any preference to turn back or move in the same direction.

3.5.4 Fractional Gaussian Noise

The stationary sequence $x(nT)$ obtained by sampling the fractional Brownian motion process $B_H(t)$ with a sampling interval T and then calculating the first difference is known as the discrete fractional Gaussian noise. It can be represented as

$$x(nT) \triangleq B_H(nT) - B_H(nT - T) \tag{3.33}$$

As the statistical properties of the fBm do not change with scale, we set $T = 1$ to obtain the discrete fractional Gaussian noise process as

$$x(n) \triangleq B_H(n) - B_H(n - 1)$$

and it is referred to as fGn.

Some important properties of the fGn are discussed next.

The autocorrelation of the discrete fractional Gaussian noise is

$$r_x(l) = \frac{1}{2}\sigma_H^2 \left(|l - 1|^{2H} - 2|l|^{2H} + |l + 1|^{2H} \right) \tag{3.34}$$

The fGn is wide-sense stationary as the correlation depends only on the distance l between the samples. Also for $H = 1/2$, we have $r_x(l) = \delta(l)$ which implies that the fGn process is white noise.

The power spectrum of the fGn process $x(n)$ is given by

$$R_x\left(e^{j\omega}\right) = \sum_{l=-\infty}^{\infty} r_x(l) e^{-j\omega l} = \sigma_H^2 C_H \left|1 - e^{-j\omega}\right|^2 \sum_{k=-\infty}^{\infty} \frac{1}{|\omega + 2\pi k|^{2H+1}} \tag{3.35}$$

where $C_H = 2H\Gamma(2H)\sin(\pi H)$ is a constant. For $H = 1/2$, a flat Power Spectral Density (PSD) is obtained, which implies that for this special case the fGn process is actually white noise. The discrete fGn process is self-similar at large scales (i.e. asymptotically). This can be inferred from the fact that the autocorrelation decays hyperbolically as $|l| \to \infty$, $H \neq 1/2$ and is given by

$$r(l) \sim \sigma_H^2 H(2H - 1) |l|^{2H-2} \tag{3.36}$$

for large lags. Also the PSD obeys a power law as the frequency becomes very small (or period becomes very large) i.e. as $|\omega| \to 0$, $H \neq 1/2$ and is given by

$$R\left(e^{j\omega}\right) \sim C_H \frac{\sigma_H^2}{|\omega|^{2H-1}} \tag{3.37}$$

The fGn process has long memory in the range $1/2 < H < 1$, since $\sum_{l=-\infty}^{\infty} r(l) = \infty$ or equivalently $R\left(e^{j\omega}\right) \to \infty$ as $|\omega| \to 0$. In this case the autocorrelation decays slowly and the frequency response is analogous to a low-pass filter. However, the process exhibits short memory for $0 < H < 1/2$, since $\sum_{l=-\infty}^{\infty} |r(l)| < \infty$ and $\sum_{l=-\infty}^{\infty} r(l) = 0$, or equivalently $R(e^{j\omega}) \to 0$ as $|\omega| \to 0$. Also, for $0 < H < 1/2$, the correlation is negative, that is, $r(l) < 0$ for $l \neq 0$ and hence the process exhibits anti-persistence. For this instance, the autocorrelation decays very quickly and the frequency response is analogous to that of a high-pass filter.

The fGn can alternatively be viewed as the output of a fractional integrator $(1/s^\beta, 0 < \beta < 1)$ driven by continuous time stationary white Gaussian noise $w(t)$ with variance σ^2. The output can be written in the following convolution form:

$$y_\beta(t) = \frac{1}{\Gamma(\beta)} \int_{-\infty}^{t} w(\tau)(t - \tau)^{\beta-1} d\tau \tag{3.38}$$

The autocorrelation of the fGn $y_\beta(t)$ is given as:

$$r(\tau) = \sigma^2 \frac{|\tau|^{2\beta-1}}{2\Gamma(2\beta) \cos \beta\pi} \tag{3.39}$$

The autocovariance function of fGn can be represented by the associated Hurst parameters also as follows:

$$R(\tau) = \frac{1}{2}\left[(\tau + 1)^{2H} + (\tau - 1)^{2H}\right] - \tau^{2H}, \quad \tau > 0 \tag{3.40}$$

3.5.5 Simulation of Fractional Brownian Motion and Fractional Gaussian Noise

Statistical self similar processes in general are difficult to simulate since no explicit mathematical formula exists for this purpose. The fBm being a special case, an explicit formula exists. But to generate a continuous version of the process an infinite memory would be required and thus only an approximate sampled version of the fBm can be generated on the computer. Thus the property of self similarity would not hold for the simulations at all scales. The fGn sequences can be generated by calculating the first order difference of the corresponding fBm sequences. Two methods for the simulation of fBm are briefly stated below.

3.5.5.1 Spectral Synthesis Method

An fBm with an index $0 < H < 1$ can be generated by this method. The method relies on the fact that once the spectral density function of the fBm can be constructed, the fBm itself can be obtained through inverse transformation.

3.5.5.2 Random Midpoint Replacement Method

This is a recursive method to generate an fBm and is based on the scaling property of the increments. For more details please refer to Samorodnitsky and Taqqu (1994).

Since the fGn is the output of a fractional integrator driven by wGn (Guglielmi 2006; Tseng et al. 2000), computer simulation of fGn can be obtained as the output of higher order continuous or discrete time realizations of fractional operators as discussed in Chap. 2. Some other popular fGn generation methods are listed below.

3.5.5.3 Fast Fractional Gaussian Noise Generator

This method is actually computes the Fast Fractional Gaussian Noise (FfGn) (Mandelbrot 1971) as an approximation of the discrete fractional Gaussian noise. FfGn can be expressed as the sum of low and high frequency components as

$$X_i = X_i^{(L)} + X_i^{(H)}$$

where, $X_i^{(L)}$ denotes the low frequency terms which is a weighted sum of Markov-Gauss processes and $X_i^{(H)}$ denotes the high frequency terms which is a Markov-Gauss process. $X_i^{(L)}$ can be defined as

$$X_i^{(L)} = \sum_{n=0}^{N} W_n^{\frac{1}{2}} M_t^{(n)}$$

where, $M_t^{(n)}$ represents the nth Markov-Gauss process with variance 1 and weight factor $W_n^{\frac{1}{2}}$ and

$$W_n = \frac{H(2H-1)}{\Gamma(3-2H)}(B^{1-H} - B^{H-1})B^{2n(H-1)} \qquad (3.41)$$

$X_i^{(H)}$ is used to compensate for the approximation leading to (3.41).

3.5.5.4 Multiple Timescale Fluctuation Approach

The weighted sum of the exponential function of the time lag leads to an approximation of the fGn autocorrelation function Eq. 3.39 on the basic time scale. An algorithm to generate fGn can be derived using this property. A wide range of numerical simulation indicate that the autocovariance function in Eq. 3.40 can be best approximated in the mean square sense by the following expression (Koutsoyiannis 2002):

$$R = 1.52 \, (H - 0.5)^{1.32} \qquad\qquad (3.42)$$

The principle of this algorithm is same as the FfGn algorithm.

3.5.5.5 Disaggregation Approach

Since the statistics of the aggregated fGn have simple expressions, an inductive algorithm can be used for generating fGn. The length of the series to be generated is $n = 2^m$ (where, m is an integer). Incase, $n > 2^m$, then the series can be generated with the next power of 2 and the excess terms over and above the required ones can be discarded. Initially, the single value $Z_1^{(n)}$ is generated from the knowledge of its variance and is then disaggregated into two variables on the time scale $n/2$, i.e. $Z_1^{n/2}$ and $Z_2^{n/2}$. The step is repeated until the series $Z_1^{(1)} \equiv X_1, \ldots, Z_n^{(1)} \equiv X_n$ is obtained. One step of the induction algorithm for fGn generation can be described as follows. Assuming that the fGn has been generated for the time scale $k \leq n$, the time series generated for the next scale $k/2$ can be described as follows. In the generation step, the higher level amount $z_i^{(k)} (1 < i < n/k)$ is disaggregated into two lower level amounts $Z_{2i-1}^{(k/2)}$ and $Z_{2i}^{(k/2)}$ in such a manner so that the following relation holds:

$$Z_{2i-1}^{(k/2)} + Z_{2i}^{(k/2)} = Z_i^{(k)} \qquad\qquad (3.43)$$

Thus, it is adequate to generate $Z_{2i-1}^{(k/2)}$ and obtain $Z_{2i}^{(k/2)}$ from Eq. 3.43. Thus at this generation step, the values of the previous lower level time steps $Z_1^{(k/2)}, \ldots, Z_{2i-2}^{(k/2)}$ and the values of the next higher level time steps $Z_{i+1}^{(k)}, \ldots, Z_{n/k}^{(k)}$ are both available. Theoretically speaking all correlations of $Z_{2i-1}^{(k/2)}$ with the preceeding lower level variables and the subsequent higher level variables are required. But a simplification to the algorithm can be made by considering the correlations of $Z_{2i-1}^{(k/2)}$ with one higher level time step behind and one ahead. Under this approximation $Z_{2i-1}^{(k/2)}$ can be generated by

$$Z_{2i-1}^{(k/2)} = a_2 Z_{2i-3}^{(k/2)} + a_1 Z_{2i-2}^{(k/2)} + b_0 Z_i^{(k)} + b_1 Z_{i+1}^{(k)} + V \qquad\qquad (3.44)$$

where, a_2, a_1, b_0 and b_1 are the parameters to be estimated and V is the innovation whose variance has also to be estimated.

3.5.6 *Fractional Levy Stable Motion and Noise*

For a self-similar process, if the probability density function of the stationary increments is $S\alpha S$ it is known as fractional Levy Stable motion (fLSm). The fLSm have
heavier tails of the stable distribution than the fBm and hence show more fluctuations
in the time series than their fBm counterparts. Thus they can be used to model signals
having greater variability and long memory.

An fLSm process $L_{H,\alpha}(t)$ can be expressed in terms of its increment process
$x_{H,\alpha}(n)$ and is known as fractional Levy Stable noise (fLSn). The fLSn is defined
by the stochastic integral

$$x_{H,\alpha}(n) = L_{H,\alpha}(n+1) - L_{H,\alpha}(n)$$

$$= C \int_{-\infty}^{n} \left[(n+1-s)^{H-1/\alpha} - (n-s)^{H-1/\alpha}\right] w_\alpha(s)\, ds \qquad (3.45)$$

where, C is a constant, α is the characteristic exponent of the $S\alpha S$ distribution, and
$w_\alpha(s)$ is white noise from an $S\alpha S$ distribution. Equation 3.45 reduces to an integral
description of fGn for $\alpha = 2$. Since, $S\alpha S$ distributions have infinite variance, the
second-order moments of the fLSm process do not exist (unlike that of fBm). For
more details please look in Adler et al. (1998). Some applications of fLSm in signal
processing can be found in Kogon and Manolakis (1996), Peng et al. (1993), Burnecki
and Weron (2004), etc.

3.6 Multi-Fractional Gaussian Noise

Various processes which exhibit LRD can be modeled more accurately by Fractional
Gaussian noise with a constant Hurst exponent rather than the conventional short
range dependent stochastic processes like ARMA, Markov or Poisson. α-stable self-
similar stochastic processes which are based on fGn can be used to model processes
with heavy tailed distributions which have LRD. However, since the Hurst parameter
is constant, the local scaling characteristic of the stochastic process may not be cap-
tured effectively in many complicated dynamic systems. Multi-fractional Gaussian
noise (mGn) and multi-fractional α-stable noise have local self similarity characteristics since the constant Hurst parameter (H) is generalized to the case where H is
substituted by a time varying Holder exponent $H(t)$ such that $0 < H(t) < 1$. These
processes are expedient in characterizing non-stationary and nonlinear dynamic
systems.

From the perspective of fractional order signals and systems, if the white Gaussian
noise (wGn) is integrated with the fractional order αth integration, this gives rise to
the fGn. The fractional Brownian motion (fBm) can be considered as the $(\alpha + 1)$th
fractional integration of wGn. The generalized fBm with time varying Holder exponent is known as multi-fractional Brownian motion or mBm (Peltier and Véhel 1995).

The mBm is represented in integral form as

$$B_{H(t)}(t) = \frac{\sigma}{\Gamma(H(t)+1/2)} \left\{ \begin{array}{l} \int_{-\infty}^{0} \left[(t-s)^{H(t)-1/2} - (-s)^{H(t)-1/2}\right] dB(s) \\ + \int_{0}^{t} (t-s)^{H(t)-1/2} dB(s) \end{array} \right\}$$

(3.46)

where, $B(s)$ is the standard Brownian motion, $\sigma^2 = \text{var}\left(B_{H(t)}(t)\right)\big|_{t=1}$ and $\text{var}(X)$ stands for the variance of X.

Thus fBm is a special case of mBm with constant Holder exponent with $H(t) = H$. Multi-fractional Gaussian noise is obtained by differentiating the mBm (Navarro et al. 2006). The mGn is a non-stationary process as the mGn, in general, does not have stationary increments. Hence the mGn provides a better model for non-stationary, nonlinear dynamic systems.

3.6.1 Concept of Variable Order Integrator

If the order of the fractional integral α is time varying (Lorenzo and Hartley 2002), then the definitions of Riemann–Liouville and Caputo fractional order operators can be extended to incorporate this as shown below.

The Riemann–Liouville definition for variable integral order is given as

$$_cD_t^{-q(t)} f(t) = \frac{1}{\Gamma(q(t))} \int_c^t (t-\tau)^{q(t)-1} f(\tau) d\tau + \psi(f, -q(t), a, c, t)$$

(3.47)

For $t > c$ and $\psi(f, -q(t), a, c, t)$ is the initialization function with $0 \leq q(t) < 1$. The Caputo definition for variable integral order is given as (Coimbra 2003)

$$D_t^{\alpha(t)} f(t) = \frac{1}{\Gamma(1-\alpha(t))} \int_0^t \frac{f'(\tau) d\tau}{(t-\tau)^{\alpha(t)}} + \frac{\left(f(0^+) - f(0^-)\right) t^{-\alpha(t)}}{\Gamma(1-\alpha(t))}$$

(3.48)

for $0 < \alpha(t) < 1$

Similar to the relationship between wGn, fGn and fBm the relationship between wGn, mGn and mBm can be established by replacing the constant Hurst exponent by local Holder exponent $H(t)$, and replacing the constant α by $\alpha(t)$. Thus the mGn can be obtained from the output of variable-order fractional integrator with wGn as the input (Sheng et al. 2011). The mGn $Y_{H(t)}(t)$ can be described as

$$Y_{H(t)}(t) = {}_0D_t^{-\alpha(t)} \omega(t) = \frac{1}{\Gamma(\alpha(t))} \int_{-\infty}^t (t-\tau)^{\alpha(t)-1} \omega(\tau) d\tau$$

(3.49)

where, $0 < \alpha(t) < 1/2$ and $H(t) = 1/2 + \alpha(t)$.

According to the definition of the mGn, mBm is the integration of mGn, so the mBm is the $[\alpha(t) + 1]$ th integration of wGn. Assuming $\omega(t) = 0$ when $t < 0$, the mBm can be described as:

$$B_{H(t)}(t) = {}_{-\infty}D_t^{-1-\alpha(t)}\omega(t)$$

$$= \frac{1}{\Gamma(H(t) + 1/2)} \int_0^t (t - \tau)^{H(t)-1/2}\,\omega(\tau)\,d\tau \qquad (3.50)$$

where, $1/2 < H(t) < 1$, and $\omega(t)$ is wGn.

The concept of variable order integrator can be extended to random order fractional derivative and integral, where the fractional order contains a noise term and useful for characterizing complex stochastic real world processes (Sun et al. 2011).

3.7 The Fractal Dimension

The fractal dimension is a statistical measure which reflects how completely a fractal appears to fill space as one zooms down to smaller scales. The concept of fractal dimension D (also known as the Haussdorff dimension) has a logical analogue with the property of self similarity or scaling. For a D-dimensional object which is subdivided into N identical replicas of itself, the side of each replica is scaled down by the relation $r = 1/\sqrt[D]{N}$ or $Nr^D = 1$. Hence we have

$$D = \frac{\log N}{\log(1/r)} \qquad (3.51)$$

Determination of the fractal dimension can be done by the box counting technique. The total number N of enclosing boxes (or rectangles), required to cover all the identical sub-series that have been scaled down by the ratio r from the whole series, is found out. Then Eq. 3.51 is used to estimate the fractal dimension. Random variables which exhibit LRD (e.g. fBm) are fractal curves whose Haussdorff dimension D is related to the Hurst exponent as $D = 2 - H$ (Falconer 2003).

3.8 Hurst Parameter Estimators

Estimation of the self similarity index or the Hurst parameter is of prime essence in this regard. LRD occurs when the Hurst parameter lies in the interval $0.5 < H < 1$. Various methods exist for finding out a constant value of H in this range and most are based on log-linear regression between appropriate variables. In other cases, the Hurst parameter may be a function of time. This might happen when the process has self similar characteristics but the nature of self similarity changes over time. The local Hurst index $H(t)$ contains information about the behavior of the process and has different method of evaluation.

3.8.1 Estimation of the Constant Hurst Index H in the Interval (0.5, 1)

Let, $Y_0, Y_1, \ldots, Y_{N-1}$ be a fBm sample observed at the time points $t_0 = 0, t_1 = 1/N, \ldots, t_{N-1} = (N-1)/N$ respectively.

Let, $X_k = Y_{k+1} - Y_k, k = 0, 1, \ldots, N-2$, then $\mathrm{Var}\,(X_k) = N^{-2H}$ for $k = 0, 1, \ldots, N-2$.

3.8.1.1 Time Domain Method (Aggregated Variance Method)

In the aggregated variance method the series $\{X_k\}$ is broken down into several subsequences of size m, where $m \in Z_+$. The aggregated process $\left\{X_k^{(m)}\right\}$ is defined by $X_k^{(m)} = \frac{1}{m}(X_{km} + \cdots + X_{(k+1)m-1}) \; \forall k \in Z_+$.

Due to self similarity of the process Y the process $X^{(m)} = \left\{X_k^{(m)}, k \geq 0\right\}$ has the same finite dimensional distribution as the process $m^{H-1}X$, where $X = \{X_k, k \geq 0\}$. Thus, $\mathrm{Var}(X_k^{(m)}) = m^{2H-2}\mathrm{Var}\,(X_k) = m^{2H-2}N^{-2H}$ is the same for every $k \geq 0$. Let, $M = \lfloor N/m \rfloor$. An estimator for $\mathrm{Var}(X_k^{(m)})$ is

$$\widehat{\mathrm{Var}(X_k^{(m)})} = \frac{1}{M} \sum_{i=0}^{M-1} \left(X_i^{(m)} - \overline{X^{(m)}}\right)^2 \tag{3.52}$$

where, $\overline{X^{(m)}}$ denotes the average of the sample $X^{(m)}$ and is given as

$$\overline{X^{(m)}} = \frac{1}{M} \sum_{i=0}^{M-1} X_i^{(m)} \tag{3.53}$$

The estimator of H is obtained by plotting $\widehat{\mathrm{Var}(X_k^{(m)})}$ along the Y axis and m along the X axis in a log–log graph. If the estimates of the variances are the same as the actual values then the points would lie on a line with a slope $2H - 2$. In practical cases, a line is fitted along these data points and the Hurst parameter is obtained from the slope of the fitted line.

3.8.1.2 Frequency Domain Method (the Periodogram Method)

The periodogram is defined by

$$I(\omega) = \frac{1}{N} \left| \sum_{k=0}^{N-1} (X_k - \overline{X}) e^{ik\omega} \right|^2 \tag{3.54}$$

where, ω is the frequency, $i = \sqrt{-1}$, $\overline{X}$ is the sample average. Like the spectral density, the periodogram is symmetric around zero. The periodogram is an asymptotic unbiased estimator of the spectral density f, i.e.

$$\lim_{N \to \infty} E[I(\omega)] = f(\omega) \tag{3.55}$$

The values of $I(\omega_k)$ is computed for $k = 1, \ldots, N$, where $\omega_k = \pi k/N$.

The values of the periodogram at the sample frequencies ω_k are plotted on a log–log scale. An estimate of the Hurst parameter H can be obtained from the slope of the curve which is theoretically equal to $1 - 2H$. Many other methods like discrete variations method, Higuchi method, R/S analysis, method of variance of regression residuals, etc. can be used for the estimation of the Hurst parameter. For details please refer to Rao (2010).

3.8.2 Estimation of the Time Varying Hurst Parameter for a Locally Self Similar Process

For a locally self similar process, the time varying Hurst parameter can be estimated by using wavelets as outlined below.

Let $\{Y(t), -\infty \leq t < \infty\}$ be a stochastic process with $E[Y(t)] = 0$ for every $t \geq 0$ and with covariance

$$R_t(u_1, u_2) = E[Y(t + u_1) Y(t + u_2)] \tag{3.56}$$

This process is said to be locally self-similar if

$$R_t(u_1, u_2) = R_t(0, 0) - C(t) |u_1 - u_2|^{2H(t)} (1 + O(1)) \tag{3.57}$$

As $|u_1| + |u_2| \to 0$, for every $t \geq 0$, where $C(t) > 0$. The function $H(t)$ is known as the local scaling exponent function. An example of such type of process is the generalized fractional Brownian motion. Let ψ denote the Daubechies mother wavelet and let $\widehat{Y}_a(t)$ denote the wavelet transform of the locally self-similar process Y corresponding to the scale a and location t. Then,

$$\widehat{Y}_a(t) = \sqrt{a} \int_{-\infty}^{\infty} \psi(x) Y(t + ax) \, dx \tag{3.58}$$

also

$$E\left|\widehat{Y}_a(t)\right|^2 \simeq a \int\limits_{-\infty}^{\infty} \int\limits_{-\infty}^{\infty} \psi(x)\,\psi(y)\left[R_t(0,0) - C(t)\,|ax - ay|^{2H(t)}\right]dxdy \quad = C_1 a^{1+2H(t)}$$

$$(3.59)$$

where

$$C_1 = C(t) \int\limits_{-\infty}^{\infty} \int\limits_{-\infty}^{\infty} |x - y|^{2H(t)}\,\psi(x)\,\psi(y)dxdy \qquad (3.60)$$

Let,

$$y_t(a) = \log\left(\left|\widehat{Y}_a(t)\right|^2\right), C_2 = E\left[\log\left(\left|\widehat{Y}_a(t)\right|^2 / E\left|\widehat{Y}_a(t)\right|^2\right)\right], \text{ and}$$

$$\varepsilon_t(a) = \log\left(\left|\widehat{Y}_a(t)\right|^2 / E\left|\widehat{Y}_a(t)\right|^2\right) - C_2.$$

Then,

$$y_t(a) = C_2 + \log\left(E\left|\widehat{Y}_a(t)\right|^2\right) + \varepsilon_t(a) \qquad (3.61)$$

A regression model can be derived from the above for small scale a as

$$y_t(a) \simeq c + (2H(t) + 1)\log a + \varepsilon_t(a) \qquad (3.62)$$

where $c = \log C_1 + C_2$.

$H(t)$ can be estimated from the above regression model by the method of least squares as

$$\widehat{H}_k(t) = \frac{1}{2}\left[\frac{\sum_{j=1}^{k}(x_j - \bar{x})(y_j - \bar{y})}{\sum_{j=1}^{k}(x_j - \bar{x})^2} - 1\right] \qquad (3.63)$$

where, $x_j = \log a_j$ and $y_j = y_t(a_j)$ for $j = 1, \ldots, k$, and $\bar{x} = \frac{1}{k}\sum_{j=1}^{k} x_j$ and $\bar{y} = \frac{1}{k}\sum_{j=1}^{k} y_j$.

Assuming Y is a Gaussian process and the covariance function is of the form as in (3.57)

$$\widehat{H}_k(t) \xrightarrow{p} H(t) \text{ as } k \to \infty$$

Details of the method along with a more rigorous proof can be found in Wang et al. (2001).

References

Adler, R.J., Feldman, R.E., Taqqu, M.S.: A Practical Guide to Heavy Tails: Statistical Techniques and Applications. Birkhauser, New York (1998)

Burnecki, K., Weron, A.: Levy stable processes. From stationary to self-similar dynamics and back. An application to finance. Acta Physica Polonica Series B **35**(4), 1343–1358 (2004)

Coimbra, C.F.M.: Mechanics with variable order differential operators. Annalen der Physik **12** (11–12), 692–703 (2003)

Doukhan, P., Oppenheim, G., Taqqu, M.S.: Theory and Applications of Long-Range Dependence. Birkhauser, New York (2003)

Falconer, K.J.: Fractal Geometry: Mathematical Foundations and Applications. Wiley, New York (2003)

Grossglauser, M., Bolot, J-.C.: On the relevance of long-range dependence in network traffic. IEEE/ACM Trans. Netw. **7**(5), 629–640 (1999). doi:10.1109/90.803379

Guglielmi, M.: 1/f[alpha] signal synthesis with precision control. Signal Process. **86**(10), 2548–2553 (2006). doi:10.1016/j.sigpro.2006.02.012

Hosking, J.R.M: Fractional Differencing. Biometrika **68**(1), 165–176 (1981)

Karmeshu, Krishnamachari, A.: Sequence variability and long-range dependence in DNA: an information theoretic perspective. In: Pal, N., Kasabov, N., Mudi, R., Pal, S., Parui, S. (eds.) Neural Information Processing. Lecture Notes in Computer Science, vol. 3316, pp. 1354–1361. Springer, Berlin / Heidelberg (2004)

Kogon, S.M., Manolakis, D.G.: Signal modeling with self-similar α-stable processes: the fractional Levy stable motion model. IEEE Trans. Signal Process. **44**(4), 1006–1010 (1996)

Koutsoyiannis, D.: The Hurst phenomenon and fractional Gaussian noise made easy/Le phénomène de Hurst et le bruit fractionnel gaussien rendus faciles dans leur utilisation. Hydrol. Sci. J. **47**(4), 573–595 (2002)

Lorenzo, C.F., Hartley, T.T.: Variable order and distributed order fractional operators. Nonlinear Dyn. **29**(1), 57–98 (2002)

Magin, R., Ortigueira, M.D., Podlubny, I., Trujillo, J.: On the fractional signals and systems. Signal Process. **91**(3), 350–371 (2011). doi:10.1016/j.sigpro.2010.08.003

Mandelbrot, B.B.: A fast fractional Gaussian noise generator. Water Resour. Res. **7**(3), 543–553 (1971)

Mandelbrot, B.B.: The fractal geometry of nature. Wh Freeman, New York (1983)

Mandelbrot, B.B., VanNess, J.W.: Fractional Brownian motions, fractional noises and applications. SIAM Rev. **10**(4), 422–437 (1968)

Manolakis, D.G., Ingle, V.K., Kogon, S.M., Ebrary, I.: Statistical and Adaptive Signal Processing: Spectral Estimation, Signal Modeling, Adaptive Filtering, and Array Processing. Artech House, London (2005)

Montanari, A., Toth, E.: Calibration of hydrological models in the spectral domain: an opportunity for scarcely gauged basins. Water Resour. Res. **43**(5), W05434 (2007)

Navarro, Jr., R., Tamangan, R., Guba-Natan, N., Ramos, E., Guzman, A.: The identification of long memory process in the Asean-4 stock markets by fractional and multifractional Brownian motion. Philipp. Stat. **55**(1–2), 65–83 (2006)

Nolan, J.: Stable Distributions: Models for Heavy-Tailed Data. Birkhauser, New York (2003)

Peltier, R.F., Véhel, J.L.: Multifractional Brownian motion: definition and preliminary results. Rapport de Recherche-Institut National de Recherche En Informatique Et En automatique (1995)

Peng, C.K., Mietus, J., Hausdorff, J., Havlin, S., Stanley, H.E., Goldberger, A.: Long-range anticorrelations and non-Gaussian behavior of the heartbeat. Phys. Rev. Lett. **70**(9), 1343–1346 (1993)

Rao, B.L.S.P.: Statistical Inference for Fractional Diffusion Processes. Wiley, New York (2010)

Samorodnitsky, G., Taqqu, M.S.: Stable Non-Gaussian Random Processes. Chapman & Hall, New York (1994)

Sheng, H., Sun, H., Chen, Y., Qiu, T.: Synthesis of multifractional Gaussian noises based on variable-order fractional operators. Signal Process. **91**(7), 1645–1650 (2011). doi:10.1016/j.sigpro.2011.01.010

Sun, H., Chen, Y., Chen, W.: Random-order fractional differential equation models. Signal Process. **91**(3), 525–530 (2011). doi:10.1016/j.sigpro.2010.01.027

Tseng, C-.C., Pei, S-.C., Hsia, S-.C.: Computation of fractional derivatives using Fourier transform and digital FIR differentiator. Signal Process. **80**(1), 151–159 (2000). doi:10.1016/s0165-1684(99)00118-8

Varotsos, C., Kirk-Davidoff, D.: Long-memory processes in ozone and temperature variations at the region 60 S? 60 N. Atmos. Chem. Phys. **6**(12), 4093–4100 (2006)

Wang, Y., Cavanaugh, J.E., Song, C.: Self-similarity index estimation via wavelets for locally self-similar processes. J. Stat. Plann. Inference **99**(1), 91–110 (2001). doi:10.1016/s0378-3758(01)00075-1

Chapter 4
Fractional Order Integral Transforms

Abstract This chapter introduces different transforms commonly encountered in signal processing applications. The importance of the Fractional Fourier Transform (FrFT) is highlighted and its interpretation and relation to various other transforms are also presented. The concept of filtering in the fractional domain and various other applications of signal processing using FrFT are also discussed. Other transforms like the fractional Sine, Cosine and the Hartley transforms which are obtained by suitable modification of the FrFT kernel is discussed. The fractional B-Splines and fractional wavelet transforms are also briefly introduced.

Keywords Fractional Fourier Transform · Fractional B-splines · Fractional Wavelet Transform · FrFT Based Filtering

Integral transforms techniques have become popular tools for scientists and engineers to solve real world mathematical problems. There are several integral transform techniques available like Fourier Transform, Laplace Transform, Hankel Transform, Mellin Transform, Hilbert Transform, Sine and Cosine Transform, Z Transform, Legendre Transform, Laguerre Transform, Radon Transform, Wavelet Transform, etc. Exhaustive analysis on various integral transforms and their applications can be found in Debnath and Bhatta (2007) to solve fractional calculus related problems and in Poularikas (1999) for theory of signal processing. In this chapter, few fractional order integral transforms are introduced which have been found to have advantages in signal processing over conventional integral transforms.

4.1 Fractional Fourier Transform as a Generalization of the Standard Integer Order Fourier Transform

The standard integer order Fourier Transform maps a signal from its time domain representation to its frequency domain allowing one to visualize quantitatively the different proportions of various frequencies present in the signal. Fourier Transforms are used widely in different fields of physics and engineering.

However, the transform does not give information about the local time–frequency characteristics of the signal which are especially important in the context of non-stationary signal analysis. Various other mathematical tools like the Wavelet transform, Gabor transform, Wigner–Ville distribution, etc. have been used to analyze such kind of signals. The Fractional Fourier transform (FrFT) which is a generalization of the integer order Fourier transform can also be used in this context. The repeated Fourier transform $\mathcal{F}\mathcal{F}f(x)$ yields $f(-x)$ and applying it four times $\mathcal{F}\mathcal{F}\mathcal{F}\mathcal{F}f(x)$ yields the original function $f(x)$. This can be expressed in the generalized notation $\mathcal{F}_a$ where "a" is an integer representing the number of repeated application of the transform on the input signal. The definition of the Fourier transform is suitably modified so that "a" can take real non-integer values and is known as the Fractional Fourier Transform (FrFT) (Ozaktas et al. 2001; Ozaktas and Aytür 1995). It can be shown that the application of the FrFT produces a signal representation that can be considered as a rotated time–frequency representation of the signal. The rotation angle α is related to "a" by $\alpha = a\pi/2$.

The FrFT is a linear operator and is defined as:

$$X_\alpha(u) = \mathcal{F}_\alpha(x(t)) = \int\limits_{-\infty}^{\infty} x(t) K_\alpha(t, u) dt \tag{4.1}$$

where, $K_\alpha(t, u)$ represents the Kernel function defined as

$$K_\alpha(t, u) = \begin{cases} \sqrt{\dfrac{1 - j \cot \alpha}{2}} \\ \times e^{j(u^2/2)\cot\alpha} e^{j(t^2/2)\cot\alpha - jut\cos ec\alpha} & \text{if } \alpha \neq 2\pi \\ \delta(t - u) & \text{if } \alpha = 2n\pi \\ \delta(t + ut) & \text{if } \alpha + \pi = 2n\pi \end{cases} \tag{4.2}$$

and $\delta(t)$ represents the Dirac's delta function.

When $\alpha = \pi/2$, the standard Fourier transform can be obtained from the above definition. For $\alpha = 0$, the signal itself is obtained and for $0 < \alpha < \pi/2$ the rotated time–frequency representation of the signal is obtained.

The FrFT exhibits the following properties (Sejdic et al. 2011; Bultheel and Martinez 2002):

1. *Linearity.* $\mathcal{F}_\alpha \left(\sum_k c_k x_k(t) \right) = \sum_k c_k \mathcal{F}_\alpha (x_k(t))$
2. *Commutativity.* $\mathcal{F}_{\alpha_1} \mathcal{F}_{\alpha_2} = \mathcal{F}_{\alpha_2} \mathcal{F}_{\alpha_1}$
3. *Associativity.* $\left(\mathcal{F}_{\alpha_3} \mathcal{F}_{\alpha_2} \right) \mathcal{F}_{\alpha_1} = \mathcal{F}_{\alpha_3} \left(\mathcal{F}_{\alpha_2} \mathcal{F}_{\alpha_1} \right)$
4. Repeated application of the FrFT of different orders is equal to a single application of the FrFT whose order is equal to the sum of individual orders i.e. $\mathcal{F}_{\alpha_1} \mathcal{F}_{\alpha_2} = \mathcal{F}_{\alpha_1 + \alpha_2}$.
5. The inverse FrFT can be obtained by applying $\mathcal{F}_{-\alpha}$ to the transformed signal i.e. $\mathcal{F}_\alpha \mathcal{F}_{-\alpha} = \mathcal{F}_0$ where, $\mathcal{F}_0$ is the identity operator, i.e. $\mathcal{F}_0(x(t)) = x(t)$.
6. FrFT satisfies the Parseval's theorem, i.e. $\langle x(t), y(t) \rangle = \langle X_\alpha(u), Y_\alpha(u) \rangle$.

Table 4.1 FrFT of some commonly used operations

Signal	FrFT
$x(t - \tau)$	$\exp\left(j\left(\tau^2/2\right)\sin\alpha\cos\alpha - ju\tau\sin\alpha\right) \times X_\alpha(u - \tau\cos\alpha)$
$x(t)\exp(jvt)$	$\exp\left(-jv^2(\sin\alpha\cos\alpha)/2 + jv\cos\alpha\right) \times X_\alpha(u - v\sin\alpha)$
$x(t)t$	$u\cos\alpha X_\alpha(u) + j\sin\alpha X'_\alpha(u)$
$x(t)/t$	$-j\sec\alpha\exp\left(j\left(u^2/2\right)\cot\alpha\right) \times \int_{-\infty}^{u} x(z)\exp\left(-j\left(z^2/2\right)\cot\alpha\right)dz$
$x(ct)$	$\sqrt{\frac{1-j\cot\alpha}{c^2-j\cot\alpha}} X_\beta\left(\frac{u\sin\beta}{c\sin\alpha}\right)\exp\left(j\left(u^2/2\right)\cot\alpha\left(1 - \left(\cos^2\beta/\cos^2\alpha\right)\right)\right)$
	where $\cot\beta = \cot\alpha/c^2$
$x'(t)$	$X'_\alpha(u)\cos\alpha + ju\sin\alpha X_\alpha(u)$
$\int_b^t x\left(t'\right)dt'$	$\sec\alpha\exp\left(-j\left(u^2/2\right)\tan\alpha\right) \times \int_b^u X_\alpha(z)\exp\left(j\left(z^2/2\right)\tan\alpha\right)dz$
	if $\alpha - \pi/2$ is not a multiple of π

7. The αth order FrFT shares the same eigen functions as the Fourier Transform, but its eigen-values are the αth root of the eigen-values of the standard Fourier transform (Namias 1980).

For the proofs of these properties and other discussions please refer to Mendlovic and Ozaktas (1993), Ozaktas and Mendlovic (1993) and Almeida (1994).

4.1.1 Some Basic Properties of FrFT

The FrFT of a signal under scaling, time shift, differentiation etc. which are commonly encountered operations in signal processing applications are listed in Table 4.1 (Sejdic et al. 2011).

4.1.2 FrFT of Some Commonly Used Functions

FrFT of some basic functions which can serve as building blocks for other complicated functions with addition, multiplication or convolution are listed in Table 4.2 (Sejdic et al. 2011).

4.1.3 Relation of the FrFT with Other Transforms

Since the FrFT can be interpreted as the rotation in the time–frequency plane of the signal, it can be related to other important transforms in the time–frequency plane (Bultheel and Martinez 2002). Some of these transforms and their relation to FrFT are discussed next.

Table 4.2 FrFT of some commonly used functions

Signal	FrFT
1	$\sqrt{1 + j \tan \alpha} \exp\left(-j\left(u^2/2\right)\tan\alpha\right)$
$\delta(t - \tau)$	$\sqrt{\frac{1 - j\cot\alpha}{2\pi}} \exp\left(\begin{array}{c} j\left(\left(\tau^2 + u^2\right)/2\right)\cot\alpha \\ -ju\tau\csc\alpha \end{array}\right)$
$\exp(-t^2/2)$	$\exp(-u^2/2)$
$\exp(j\eta t)$	$\sqrt{1 + j\tan\alpha} \exp\left(\begin{array}{c} j\left(\left(\eta^2 + u^2\right)/2\right)\tan\alpha \\ -ju\eta\sec\alpha \end{array}\right)$
$\exp\left(c\left(-t^2/2\right)\right)$	$\sqrt{\frac{1 - j\cot\alpha}{c - j\cot\alpha}} \exp\left(\frac{j(u^2/2)(c^2-1)\cot\alpha}{(c^2+\cot^2\alpha)}\right) \exp\left(-c(u^2/2)\csc^2\alpha/\left(c^2 + \cot^2\alpha\right)\right)$

4.1.3.1 Wigner Distribution

For a signal f, the Wigner distribution of the signal Wf, is given as,

$$(Wf)(t, u) = \frac{1}{\sqrt{2\pi}} \int\limits_{-\infty}^{\infty} f(t + z/2)\overline{f(t - z/2)}e^{-iuz}dz \qquad (4.3)$$

This transform in a way reflects the energy distribution of the signal in the time frequency plane. Putting $f_1 = Ff$ in (4.3) we have

$$\int\limits_{-\infty}^{\infty} (Wf)(t, u)\,du = |f(t)|^2 \qquad (4.4)$$

and

$$\int\limits_{-\infty}^{\infty} (Wf)(t, u)\,dt = |f_1(u)|^2 \qquad (4.5)$$

so that,

$$\frac{1}{\sqrt{2\pi}} \int\limits_{-\infty}^{\infty}\int\limits_{-\infty}^{\infty} (Wf)(t, u)\,du\,dt = \|f\|^2 = \|f_1\|^2 \qquad (4.6)$$

which is the energy of the signal f.

The Wigner distribution of a signal f and the Wigner distribution of its FrFT are related by a rotation over an angle $-\alpha$, i.e.

$$(Wf_a)(t, u) = R_{-a}(Wf)(t, u) \qquad (4.7)$$

where, $\alpha = a\pi/2$, $f_a = F_a f$ and R_{-a} stands for a clockwise rotation of the variables (t, u) over an angle α.

In other words,

$$R_a(\mathcal{W}f_a)(t, u) = (\mathcal{W}f_a)(t_a, u_a) = (\mathcal{W}f)(t, u) \tag{4.8}$$

with $(t_a, u_a) = R_a(t, u)$.

4.1.3.2 Ambiguity Function

The ambiguity function by definition is similar to the Wigner distribution with the integral over the other variable.

$$(\mathcal{A}f)(t, u) = \frac{1}{\sqrt{2\pi}} \int\limits_{-\infty}^{\infty} f(z + t/2)\overline{f(z - t/2)}e^{-izu}dz \tag{4.9}$$

It is related to the correlation of the signal. $(\mathcal{A}f)(t, 0)$ is the autocorrelation function of f and $(\mathcal{A}f)(0, u)$ is the autocorrelation function of $f_1 = \mathcal{F}f$. Similar to the Wigner distribution, the following relation holds for the ambiguity function.

$$R_a(\mathcal{A}f_a)(t, u) = (\mathcal{A}f_a)(t_a, u_a) = (\mathcal{A}f)(t, u) \tag{4.10}$$

4.1.4 Two Dimensional Transform

The generalized one dimensional FrFT is defined as

$$\mathcal{F}_\alpha(f)(\xi) = \int\limits_{-\infty}^{\infty} K_a(\xi, x)f(x)dx \tag{4.11}$$

where, $K_a(\xi, x)$ is the kernel function.

The simplest separable generalization of the FrFT in two-dimensions is given by

$$(\mathcal{F}_c f)(\xi) = (\mathcal{F}_{a,b}f)(\xi, \eta) = \int\limits_{-\infty}^{\infty}\int\limits_{-\infty}^{\infty} K_{a,b}(\xi, \eta; x, y)f(x, y)dxdy \tag{4.12}$$

where, $K_{a,b}(\xi, \eta; x, y) = K_a(\xi, x)K_b(\eta, y)$

For the 2D-FrFT there are two angles of rotation:

$\alpha = a\pi/2$ and $\beta = b\pi/2$.

Putting one of these angles zero, results in the reduction of (4.12) to (4.11), i.e. to the one dimensional FrFT.

The FrFT can be extended to higher dimensions as

$$(\mathcal{F}_{a_1,\dots,a_n} f)(\xi_1,\dots,\xi_n)$$

$$= \int\limits_{-\infty}^{\infty} \cdots \int\limits_{-\infty}^{\infty} K_{a_1,\cdots,a_n}(\xi_1,\dots,\xi_n; x_1,\dots,x_n) f(x_1,\dots,x_n) dx_1 \dots dx_n \quad (4.13)$$

4.1.5 Discrete FrFT and its Computation

Practical realization of the FrFT requires its numerical computation (Marinho and Bernardo 1998). Since the FrFTs have quadratic complex exponential kernels, very fast oscillations are induced and numerical integration requires very fast sampling times. This results in increased computational time and memory along with larger numerical errors. Various other techniques have been used to circumvent this problem with varying degrees of accuracy and computational loads. The direct sampling of the FrFT to obtain its discrete version is mostly a simple application of the Shannon sampling theorem (Zayed and García 1999; Candan and Ozaktas 2003; Tao et al. 2008a,b). But some important properties like unitarity, reversibility, additivity, etc. is lost, thus limiting its application in many cases. The linear combination type DFrFT (Dickinson and Steiglitz 1982) is another realization which is based on linearly combining the ordinary Fourier operators raised to different powers. Its advantage is that it is an easy implementation of the Fourier operators, but it is altogether another distinct definition of the FrFT and the transformed signal does not correspond to the continuous FrFT. Another way of obtaining the DFrFT is based on the concept of eigen-value decomposition (Pei et al. 1999). The signal obtained matches the continuous FrFT and retains some important property of the FrFT. The main disadvantage of this method is that it cannot be written in a closed form and has higher computational costs. Various other approaches for computing DFrFT for e.g. using Quadratic Phase Transform (Yeh and Pei 2003; Ikram et al. 1997) and others (Xia 2000; Bi et al. 2006; Ju and Bi 2007; Djurovic et al. 2001) have also been used.

4.1.6 Filtering and Noise Removal in Fractional Domain

Filtering, which refers to recovering the desired signal from a measurement corrupted by noise, has major applications in signal processing. The concept of the digital filter design which are employed for filtering purpose may be extended to the fractional domain with additional advantages (Tao et al. 2006; Kutay et al. 1997).

The conventional filter can be written as

$$x_0(t) = \int\limits_{-\infty}^{\infty} h(t-\tau)x_i(\tau)d\tau \quad (4.14)$$

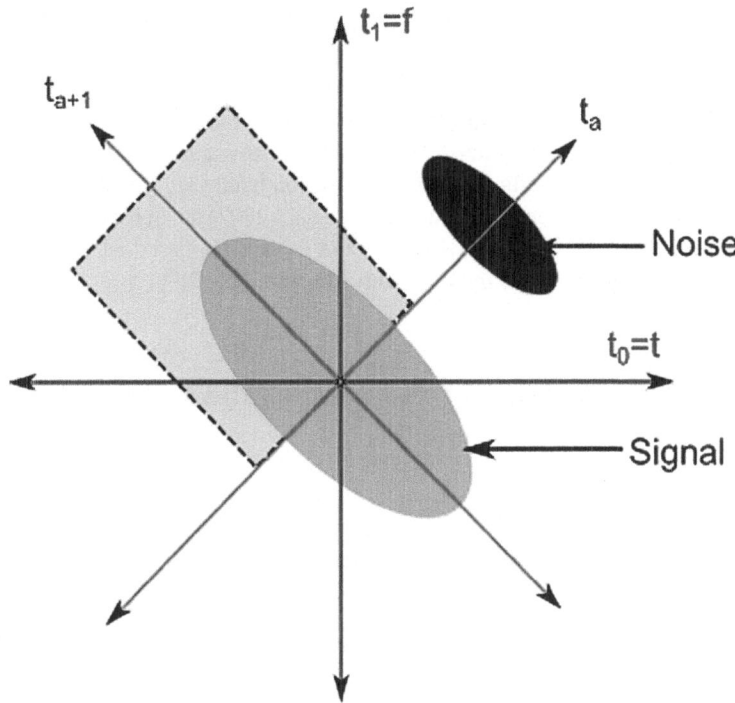

Fig. 4.1 Concept of filtering with FrFT

where $x_i(t)$, $x_0(t)$ and $h(t)$ refer to input signal, output signal and the impulse response of the filter respectively.

Rewriting (4.14) in the form

$$x_0(t) = \frac{1}{\sqrt{2\pi}} \mathcal{F}_{-1} \left(\mathcal{F}(x_i(t)) H(\omega) \right) \qquad (4.15)$$

where $H(\omega) = \mathcal{F}(h(t))$,
the fractional filter can be written as a generalization of the conventional filter as

$$x_0(t) = \mathcal{F}_{-\alpha} \{ \mathcal{F}_{\alpha} \{ x_i(t) \} \cdot H_{\alpha}(u) \} \qquad (4.16)$$

where $H_{\alpha}(u) = \mathcal{F}_{\alpha} \{ h(t) \}$.

Since performing the αth order Fractional Fourier Transform operation is equivalent to rotating the Wigner distribution by an angle $\phi = \alpha\pi/2$ in the clockwise direction, a fractional domain can be found where the signal and noise do not overlap. Thus we can do the FrFT (rotate the Wigner distribution) by a proper choice of rotation angle and then filter out the noise in the fractional domain. The concept is illustrated in the schematic below (Fig. 4.1). It is obvious that conventional filtering in the frequency domain would not be able to filter out the noise as the signal and the noise overlap in this domain.

Different window based filtering techniques, commonly used in traditional digital signal processing have been extended for FrFT based filter design. This design technique has the advantage that the additional rotation parameter of the FrFT can be used to control the main lobe width, minimum stop band attenuation, etc. The method has been applied to the generalized "Hamming" window and the Dirichlet window (Kumar et al. 2011). A method of tuning transition bandwidth of FIR filters using FrFT have been found expedient as in Sharma et al. (2007) and have been shown for two different types of window, viz. Kaiser and PC6. FrFT based adaptive filtering techniques has been proposed in Durak and Aldirmaz (2010).

4.1.7 FrFT Applications in Other Areas of Signal Processing

4.1.7.1 Watermarking

Digital watermarking is a process of embedding information in a digital signal for various purposes like visible identification (e.g., company logo on a video), steganography (writing hidden messages), copyright protection of digital multimedia data, etc. For multimedia data it is of advantage if more watermarks can be embedded in it with almost negligible aberration from the original version. FrFT watermarks with different angles have a small correlation among themselves and also can be detected only if the angle under which the watermark is embedded is known. Due to these properties the FrFT domain offers more flexibility for the purpose of embedding more number of watermarks in the signal. Exploiting the two angles of the 2D FrFT, more different watermarks can be hidden in images than that in the standard DFT based domains. For more details please refer to Djurovic et al. (2001), Savelonas and Chountasis (2010), Guo et al. (2007) and Simitopoulos et al. (2003).

4.1.7.2 Communication

DFT based techniques are conventionally used for designing multicarrier (MC) systems. The Orthogonal Frequency Division Multiplexing (OFDM) is one of the main representatives of this technique. OFDM has been used in applications like broad band internet access, wireless networking, digital TV, audio broadcasting, etc. OFDM can cope with severe channel conditions and is thus advantageous over single carrier schemes. But for channels with selectivity in both time and frequency domains (i.e. doubly selective channels), OFDM fails to give acceptable results. Discrete FrFT can be implemented over DFT in these cases. For channels having rapid variations with available line-of-sight component, there is a remarkable improvement in performance. This is due to the fact that the time–frequency plane can be rotated so as to compensate the undesired modulation of the signals, produced by these rapid variations. For further reading please refer to Martone (2001), Khanna and Saxena (2009) and Erseghe et al. (2005).

4.1.7.3 Compression

Data compression is used for encoding information using smaller number of bits than the original representation. DFrFT and Set Partitioning in Hierarchical Tree (SPIHT) have been used for data compression to achieve a high signal compression ratio. Also the comparison of this technique with the Discrete Cosine Transform (DCT) methodology shows that this has a higher quality of the reconstructed signal and better percentage RMS difference. Further details can be found in Yetik et al. (2001) and Vijaya and Bhat (2006).

4.1.7.4 Cryptography

Cryptography finds numerous applications in computer passwords, electronic banking, online trading, etc. FrFT has been used successfully in cryptographic applications. For encrypting the plain text, parameters like angle of rotation, time exponent, phase, sampling rate are chosen as the encryption key. Then a suitably modified FrFT kernel for the given encryption key is used to be multiplied with the input signal to produce the ciphertext. The reverse method is used at the receiving end to convert the ciphertext back into plain text. More details on the application of FrFT in cryptography can be found in Youssef (2008), Tao et al. (2008b) and Ran et al. (2009).

4.1.7.5 Pattern Recognition

The presence of the rotational angle in FrFT gives additional flexibility in pattern recognition systems. FrFT has been used as a pre-processor for a neural network resulting in significant reduction in classification and localization errors. The FrFT increases the computational cost only in the training phase of the neural network and does not affect the cost during operation. More details can be found in Barshan and Ayrulu (2002) and Lohmann et al. (1996b).

4.1.7.6 Fractal Signal Processing

FrFT based estimation methods have been used to analyze the long range dependence (LRD) in time series. Hurst exponent calculated by FrFT based methods have been shown to be better than other existing ones (Chen et al. 2010) like Wavlet based methods. FrFT has also been used for finding out the important parameters of fractals. Further reading can be directed to Alieva (1996).

4.1.7.7 Miscellaneous

FrFT has been applied to solving differential equations, computer tomography (Gbur and Wolf 2001), transient motor current signature analysis (TMCSA) (Pineda-Sanchez et al. 2010; Sun et al. 2002), radar applications (Sun et al. 2002), 2D signal processing (Ali Khan et al. 2011) e.g. image processing, etc. among others.

4.2 Fractional Sine, Cosine and Hartley Transforms

The Fractional Cosine Transform (FrCT), Fractional Sine Transform (FrST) and Fractional Hartley Transform (FrHT) can be obtained by a suitable modification of the FrFT kernel $K_\alpha(t, u)$.

The fractional order Cosine transform is given by

$$C_\alpha(u) = \int_{-\infty}^{\infty} x(t)\text{Re}\,[K_\alpha(t, u)]\,dt \tag{4.17}$$

where Re[·] indicates the real part of the kernel $K_\alpha(t, u)$.

The fractional order sine transform is given by

$$S_\alpha(u) = \int_{-\infty}^{\infty} x(t)\text{Im}\,[K_\alpha(t, u)]\,dt \tag{4.18}$$

where Im[·] indicates the imaginary part of the kernel $K_\alpha(t, u)$.

The fractional order Hartley transform generalizes the fractional order sine and cosine transforms and is given by

$$H_\alpha(u) = \int_{-\infty}^{\infty} x(t)\,\{\text{Im}\,[K_\alpha(t, u)] + \text{Re}\,[K_\alpha(t, u)]\}\,dt$$

$$= C_\alpha(u) + S_\alpha(u) \tag{4.19}$$

For a real input signal, the FrCT and FrST spectrum corresponds to the real and imaginary part of the FrFT spectrum. The FrHT is the fractional extension of the Hartley Transform, which finds application in image processing and data compression. The FrHT is phase-shift variant and inherits the advantages of both the FrFT and the Hartley Transform. It can be used for efficient noise filtering and pattern recognition.

The FrHT is a real transformation but its index is not additive, i.e.

$$H_{\alpha_1+\alpha_2}(u) \neq H_{\alpha_2}\left\{H_{\alpha_1}(u)\right\} \tag{4.20}$$

The indices of the fractional sine and cosine transforms are not additive as well.

$$C_{\alpha_1+\alpha_2}(u) \neq C_{\alpha_2}\left\{C_{\alpha_1}(u)\right\} \tag{4.21}$$

$$S_{\alpha_1+\alpha_2}(u) \neq S_{\alpha_2}\left\{S_{\alpha_1}(u)\right\} \tag{4.22}$$

More details can be found in Lohmann et al. (1996a), Zhang et al. (1998) and Pei et al. (1998).

4.3 Fractional B-splines and Fractional Wavelet Transform

Wavelets are widely used for signal compression and analysis. They are very useful in detecting and characterizing singularities in signals. This is due to the property that a wavelet with n vanishing moments acts as a differentiator of order $n+1$, i.e. $\widehat{\psi}(\omega) \propto \omega^{n+1}$ as $\omega \to \infty$. But for signals having a fractal nature, or fBms, it is useful if the wavelet behaves like a fractional differentiator, i.e. $\widehat{\psi}(\omega) \propto \omega^{\alpha+1}$ as $\omega \to \infty$, for some non-integer number α (Blu and Unser 2000). A set of wavelet bases have been defined in Unser and Blu (1999), which have this property. They are derived from a scaling function known as fractional B-spline [which is so named as they interpolate the polynomial B-splines at the integers (Unser and Blu 2000)]. The fractional B-splines have similar properties of the integer order B-splines, like approximation, regularity, scaling, etc. but they do not have a finite support (Unser and Blu 2000; Panda and Dash 2006).

4.3.1 The B-spline and the Fractional B-spline

A generalized version of the Bezier curve is known as the B-Spline. Let a vector known as the knot be defined as $T = \{t_0, t_1, \ldots, t_m\}$, where T represents a non-decreasing sequence with $t_i \in [0, 1]$ and let the control points be defined as P_0, P_n. The degree is defined as $p = m - n - 1$. The knots $t_{p+1}, \ldots, t_{m-p-1}$ are called internal knots.

If the basis functional is defined as

$$N_{i,0}(t) = \begin{cases} 1 & \text{if } t_i \leq t < t_{i+1} \text{ and } t_i < t_{i+1} \\ 0 & \text{otherwise} \end{cases} \tag{4.23}$$

and

$$N_{i,p}(t) = \frac{t - t_i}{t_{i+p} - t_i} N_{i,p-1}(t) + \frac{t_{i+p+1} - t}{t_{i+p+1} - t_{i+1}} N_{i+1,p-1}(t) \tag{4.24}$$

Then the B-spline is represented by the curve

$$C(t) = \sum_{i=0}^{n} P_i N_{i,p}(t) \tag{4.25}$$

The forward fractional finite difference operator of order α is given by

$$\Delta_+^\alpha f(x) = \sum_{k=0}^{+\infty} (-1)^k \binom{\alpha}{k} f(x - k) \tag{4.26}$$

where, $\binom{\alpha}{k}$ is the binomial co-efficient.

The fractional B-splines are the localized versions of the one sided power function $(x - k)_+^\alpha := \max(x - k, 0)^\alpha$ and are defined as follows

$$\beta_+^\alpha(x) := \frac{\Delta_+^{\alpha+1} x_+^\alpha}{\Gamma(\alpha + 1)} = \frac{\sum_{k=0}^{+\infty} (-1)^k \binom{\alpha + 1}{k} (x - k)_+^\alpha}{\Gamma(\alpha + 1)} \tag{4.27}$$

where, $\Gamma(\alpha + 1) = \int_0^{+\infty} x^\alpha e^{-x} dx$ and $\alpha > -1/2$ to ensure square integrability.

These functions interpolate the usual polynomial B-splines. They are termed as causal since their support belongs to $\mathbb{R}_+$.

The fractional B-splines decrease similar to $|x|^{-\alpha-2}$ as $|x| \to \infty$. They have a fractional order of approximation $\alpha + 1$. This indicates that any polynomial of degree $\leq \lceil \alpha \rceil$ can be expressed as a linear combination of $\beta^\alpha(x-k)$, $k \in \mathbb{Z}$. Thus from this perspective they act like approximators, having a higher integer order approximation of $1 + \lceil \alpha \rceil$. The fractional B-Splines exhibit the convolution property

$$\beta_+^{\alpha_1} * \beta_+^{\alpha_2} = \beta_+^{\alpha_1 + \alpha_2} \tag{4.28}$$

The fractional B-splines are not symmetric in general except for integer values of α. The centered fractional B-splines of degree α are given by

$$\beta_*^\alpha(x) = \frac{1}{\Gamma(\alpha + 1)} \sum_{k \in \mathbb{Z}} (-1)^k \left| \begin{matrix} \alpha + 1 \\ k \end{matrix} \right| |x - k|_*^\alpha \tag{4.29}$$

where $|x|_*^\alpha$ is defined as

$$|x|_*^\alpha = \begin{cases} \dfrac{|x|^\alpha}{-2 \sin(\alpha\pi/2)}, & \alpha \text{ is not even} \\ \dfrac{x^{2n} \log x}{(-1)^{1+n}\pi}, & \alpha \text{ is even} \end{cases} \tag{4.30}$$

and $\left| \begin{matrix} x \\ k \end{matrix} \right|$ is defined by the symmetrized version of the binomial function by $\left| \begin{matrix} x \\ k \end{matrix} \right| = \binom{x}{k + \frac{x}{2}}$.

4.3.2 Fractional B-spline Wavelets

The fractional B-Spline wavelets can be defined as

$$\psi_+^\alpha \left(\frac{x}{2} \right) = \sum_{k \in \mathbb{Z}} \frac{(-1)^k}{2^\alpha} \sum_{l \in \mathbb{Z}} \binom{\alpha + 1}{l} \beta_*^{2\alpha+1} (l + k - 1) \beta_+^\alpha(x - k) \tag{4.31}$$

The fractional B-spline wavelets obey the following

$$\int\limits_{-\infty}^{\infty} x^n \psi_+^\alpha(x)dx = 0 \qquad (4.32)$$

The fractional spline wavelets behave like the fractional derivative operator of order $\alpha + 1$ for low frequencies. This is due to the fact that for ω close to 0, the non-symmetric and symmetric fractional spline wavelets are $\propto (-j\omega)^{\alpha+1}$ and $\propto |\omega|^{\alpha+1}$ respectively to the first order in ω. This is a useful property when handling $1/f^\beta$ signals, since the fractional differentiator of order β, "whiten" them (Blu and Unser 2000, 2002).

The fractional wavelet transform has been successfully applied for spectral analysis of a complex mixture as reported in Dinç and Baleanu (2006) and Dinç et al. (2010).

References

Ali Khan, N., Ahmad Taj, I., Noman Jaffri, M., Ijaz, S.: Cross-term elimination in Wigner distribution based on 2D signal processing techniques. Signal Process. **91**(3), 590–599 (2011). doi:10.1016/j.sigpro.2010.06.004

Alieva, T.: Fractional fourier transform as a tool for investigation of fractal objects. JOSA A **13**(6), 1189–1192 (1996)

Almeida, L.B.: The fractional fourier transform and time–frequency representations. Signal Process. IEEE Trans. **42**(11), 3084–3091 (1994)

Barshan, B., Ayrulu, B.: Fractional fourier transform pre-processing for neural networks and its application to object recognition. Neural Netw. **15**(1), 131–140 (2002)

Bi, G., Wei, Y., Li, G., Wan, C.: Radix-2 DIF fast algorithms for polynomial time–frequency transforms. Aerosp. Electron. Syst. IEEE Trans. **42**(4), 1540–1546 (2006)

Blu, T., Unser, M.: The fractional spline wavelet transform: definition and implementation. In: Acoustics, Speech, and Signal Processing, 2000. ICASSP '00 Proceedings. 2000 IEEE International Conference on, vol.511, pp. 512–515 (2000)

Blu, T., Unser, M.: Wavelets, fractals, and radial basis functions. Signal Process. IEEE Trans. **50**(3), 543–553 (2002)

Bultheel, A., Martinez, H.: A shattered survey of the fractional fourier transform. Report TW **337** (2002)

Candan, C., Ozaktas, H.M.: Sampling and series expansion theorems for fractional fourier and other transforms. Signal Process. **83**(11), 2455–2457 (2003)

Chen, Y., Sun, R., Zhou, A.: An improved Hurst parameter estimator based on fractional fourier transform. Telecommun. Syst. **43**(3), 197–206 (2010). doi:10.1007/s11235-009-9207-4

Debnath, L., Bhatta, D.: Integral Transforms and their Applications. CRC press, London (2007)

Dickinson, B., Steiglitz, K.: Eigenvectors and functions of the discrete Fourier transform. Acoust. Speech Signal Process. IEEE Trans. **30**(1), 25–31 (1982)

Dinç, E., Baleanu, D.: A new fractional wavelet approach for the simultaneous determination of ampicillin sodium and sulbactam sodium in a binary mixture. Spectrochimica Acta Part A: Mol. Biomol. Spectrosc. **63**(3), 631–638 (2006). doi:10.1016/j.saa.2005.06.012

Dinç, E., Demirkaya, F., Baleanu, D., Kadioglu, Y., Kadioglu, E.: New approach for simultaneous spectral analysis of a complex mixture using the fractional wavelet transform. Commun. Nonlinear Sci. Numer. Simul. **15**(4), 812–818 (2010). doi:10.1016/j.cnsns.2009.05.021

Djurovic, I., Stankovic, S., Pitas, I.: Digital watermarking in the fractional fourier transformation domain. J. Netw. Comput. Appl. **24**(2), 167 (2001)

Durak, L., Aldirmaz, S.: Adaptive fractional fourier domain filtering. Signal Process. **90**(4), 1188–1196 (2010). doi:10.1016/j.sigpro.2009.10.002

Erseghe, T., Laurenti, N., Cellini, V.: Multicarrier architecture based upon the affine fourier transform. Commun. IEEE Trans. **53**(5), 853–862 (2005)

Gbur, G., Wolf, E.: Relation between computed tomography and diffraction tomography. JOSA A **18**(9), 2132–2137 (2001)

Guo, J., Liu, Z., Liu, S.: Watermarking based on discrete fractional random transform. Opt. Commun. **272**(2), 344–348 (2007)

Ikram, M.Z., Abed-Meraim, K., Hua, Y.: Fast quadratic phase transform for estimating the parameters of multicomponent chirp signals. Digit. Signal Process. **7**(2), 127–135 (1997)

Ju, Y., Bi, G.: Generalized fast algorithms for the polynomial time–frequency transform. Signal Process. IEEE Trans. **55**(10), 4907–4915 (2007)

Khanna, R., Saxena, R.: Improved fractional fourier transform based receiver for spatial multiplexed mimo antenna systems. Wirel. Pers. Commun. **50**(4), 563–574 (2009)

Kumar, S., Singh, K., Saxena, R.: Analysis of Dirichlet and Generalized "Hamming" window functions in the fractional Fourier transform domains. Signal Process. **91**(3), 600–606 (2011). doi:10.1016/j.sigpro.2010.04.011

Kutay, A., Ozaktas, H.M., Ankan, O., Onural, L.: Optimal filtering in fractional Fourier domains. Signal Process. IEEE Trans. **45**(5), 1129–1143 (1997)

Lohmann, A.W., Mendlovic, D., Zalevsky, Z., Dorsch, R.G.: Some important fractional transformations for signal processing. Opt. Commun. **125**(1–3), 18–20 (1996a)

Lohmann, A.W., Zalevsky, Z., Mendlovic, D.: Synthesis of pattern recognition filters for fractional fourier processing. Opt. Commun. **128**(4–6), 199–204 (1996b)

Marinho, F.J., Bernardo, L.M.: Numerical calculation of fractional fourier transforms with a single fast-fourier-transform algorithm. JOSA A **15**(8), 2111–2116 (1998)

Martone, M.: A multicarrier system based on the fractional fourier transform for time–frequency-selective channels. Commun. IEEE Trans. **49**(6), 1011–1020 (2001)

Mendlovic, D., Ozaktas, H.M.: Fractional fourier transforms and their optical implementation: I. J. Opt. Soc. Am. A **10**(9), 1875–1881 (1993)

Namias, V.: The fractional order fourier transform and its application to quantum mechanics. IMA J. Appl. Math. **25**(3), 241 (1980)

Ozaktas, H.M., Aytür, O.: Fractional fourier domains. Signal Process. **46**(1), 119–124 (1995). doi:10.1016/0165-1684(95)00076-p

Ozaktas, H.M., Kutay, M.A., Zalevsky, Z.: The fractional fourier transform with applications in optics and signal processing. Wiley, New York (2001)

Ozaktas, H.M., Mendlovic, D.: Fractional fourier transforms and their optical implementation. II. J. Opt. Soc. Am. A **10**(12), 2522–2531 (1993)

Panda, R., Dash, M.: Fractional generalized splines and signal processing. Signal Process. **86**(9), 2340–2350 (2006). doi:10.1016/j.sigpro.2005.10.017

Pei, S.C., Tseng, C.C., Yeh, M.H., Jian-Jiun, D.: A new definition of continuous fractional Hartley transform. In: IEEE, vol. 1483, pp. 1485–1488 (1998)

Pei, S.C., Yeh, M.H., Tseng, C.C.: Discrete fractional fourier transform based on orthogonal projections. Signal Process. IEEE Trans. **47**(5), 1335–1348 (1999)

Pineda-Sanchez, M., Riera-Guasp, M., Antonino-Daviu, J.A., Roger-Folch, J., Perez-Cruz, J., Puche-Panadero, R.: Diagnosis of induction motor faults in the fractional fourier domain. Instrum. Meas. IEEE Trans. **59**(8), 2065–2075 (2010)

Poularikas, A.D.: The Handbook of Formulas and Tables for Signal Processing. CRC, London (1999)

Ran, Q., Zhang, H., Zhang, J., Tan, L., Ma, J.: Deficiencies of the cryptography based on multiple-parameter fractional fourier transform. Opt. Lett. **34**(11), 1729–1731 (2009)

Savelonas, M.A., Chountasis, S.: Noise-resistant watermarking in the fractional fourier domain utilizing moment-based image representation. Signal Process. **90**(8), 2521–2528 (2010)

Sejdic, E., Djurovic, I., Stankovic, L.: Fractional fourier transform as a signal processing tool: an overview of recent developments. Signal Process. **91**(6), 1351–1369 (2011). doi:10.1016/j.sigpro.2010.10.008

Sharma, S.N., Saxena, R., Saxena, S.C.: Tuning of FIR filter transition bandwidth using fractional fourier transform. Signal Process. **87**(12), 3147–3154 (2007). doi:10.1016/j.sigpro.2007.06.005

Simitopoulos, D., Koutsonanos, D.E., Strintzis, M.G.: Robust image watermarking based on generalized radon transformations. Circuits Syst. Video Technol. IEEE Trans. **13**(8), 732–745 (2003)

Sun, H.B., Liu, G.S., Gu, H., Su, W.M.: Application of the fractional fourier transform to moving target detection in airborne SAR. Aerosp. Electronic Syst. IEEE Trans. **38**(4), 1416–1424 (2002)

Tao, R., Deng, B., Wang, Y.: Research progress of the fractional fourier transform in signal processing. Sci. China Ser. F: Inf. Sci. **49**(1), 1–25 (2006)

Tao, R., Deng, B., Zhang, W.Q., Wang, Y.: Sampling and sampling rate conversion of band limited signals in the fractional fourier transform domain. Signal Process. IEEE Trans. **56**(1), 158–171 (2008a)

Tao, R., Lang, J., Wang, Y.: Optical image encryption based on the multiple-parameter fractional fourier transform. Opt. Lett. **33**(6), 581–583 (2008b)

Unser, M., Blu, T.: Fractional splines and wavelets. SIAM Rev. **42**(1), 43–67 (2000)

Unser, M.A., Blu, T.: Construction of fractional spline wavelet bases. In: Unser, M.A., Aldroubi, A., Laine, A.F. (eds.) Denver, CO, USA, pp. 422–431. (1999) SPIE

Vijaya, C., Bhat, J.: Signal compression using discrete fractional Fourier transform and set partitioning in hierarchical tree. Signal Process. **86**(8), 1976–1983 (2006)

Xia, X.G.: Discrete chirp-fourier transform and its application to chirp rate estimation. Signal Process. IEEE Trans. **48**(11), 3122–3133 (2000)

Yeh, M.H., Pei, S.C.: A method for the discrete fractional fourier transform computation. Signal Process. IEEE Trans. **51**(3), 889–891 (2003)

Yetik, I.S., Kutay, M., Ozaktas, H.M.: Image representation and compression with the fractional fourier transform. Opt. Commun. **197**(4), 275–278 (2001)

Youssef, A.M.: On the security of a cryptosystem based on multiple-parameters discrete fractional fourier transform. Signal Process. Lett. IEEE **15**, 77–78 (2008)

Zayed, A.I., García, A.G.: New sampling formulae for the fractional fourier transform. Signal Process. **77**(1), 111–114 (1999)

Zhang, Y., Gu, B.Y., Dong, B.Z., Yang, G.Z.: A new kind of windowed fractional transforms. Opt. Commun. **152**(1–3), 127–134 (1998)

Chapter 5
Fractional Order System Identification

Abstract This chapter discusses various time domain and frequency domain system identification methods for fractional order systems from practical test data. System identification is important in cases where it is difficult to obtain the model from basic governing equations and first principles, or where there is only input–output data available and the underlying phenomena are largely unknown. As is evident, fractional order models are better capable of modeling system dynamics than their integer order counterparts. Hence, identification using fractional order models is of practical interest from the system designer's point of view.

Keywords Time domain FO system identification · Frequency domain FO system identification · Levy's identification method · Vinagre's identification method

System identification is used to obtain mathematical models of any dynamical system from measured data. System identification is important in cases where it is difficult to obtain the model from basic governing equations and first principles, or where there is only input–output data available and the underlying phenomena are largely unknown. Fractional order systems have better flexibility in modeling physical phenomena than their integer order counterparts and hence are gaining importance in the research community. Fractional order system identification has been used for the estimation of the state of charge for lead acid batteries (Sabatier et al. 2006). Fractional order modeling and identification of thermal systems has been investigated in Gabano et al. (2011) and Gabano and Poinot (2011a, b). The methods for system identification can be broadly classified into time domain and frequency domain techniques. Some fractional order system identification techniques are presented next based on these classifications.

5.1 Time Domain Identification Methods

Fractional order elements are generally approximated with Oustaloup's recursive formula for pole-zero distribution. For time domain identification of fractional order models, the basic building block i.e. a FO integrator $(s^{-\gamma})$ needs to be modified with an integrator instead of a gain outside the frequency range of fitting $\omega \in [\omega_l, \omega_h]$, as suggested in Malti et al. (2007).

$$s^{-\gamma} \simeq \frac{C_0}{s}\left(\frac{1+\frac{s}{\omega_A}}{1+\frac{s}{\omega_B}}\right)^{1-\gamma} \approx \frac{C_0}{s}\prod_{k=1}^{N}\left(\frac{1+\frac{s}{\omega_k}}{1+\frac{s}{\omega_k'}}\right) \qquad (5.1)$$

where $\omega_k = \alpha\omega_k'$, $\omega_{k+1}' = \eta\omega_k'$ and

$$\gamma = 1 - \frac{\log\alpha}{\log\alpha\eta} \qquad (5.2)$$

Here, $\{\alpha, \eta\}$ are real parameters and depends on the order of FO integration γ. Increase in the order of approximation N increases the accuracy of the rational model with an undesirable increase in the model complexity. The rationalized state space model has the following form

$$\begin{aligned}\dot{x}(t) &= Ax(t) + Bu(t)\\y_{I\gamma}(t) &= Cx(t)\end{aligned} \qquad (5.3)$$

where $y_{I\gamma}(t)$ being the γth order fractional integral of input $u(t)$,

$$A = \begin{bmatrix} 1 & 0 & \cdots & \cdots & 0 \\ -\alpha & 1 & & & \vdots \\ 0 & -\alpha & 1 & & \vdots \\ \vdots & & \ddots & \ddots & 0 \\ 0 & \cdots & 0 & -\alpha & 1 \end{bmatrix}\begin{bmatrix} 0 & 0 & \cdots & \cdots & 0 \\ \omega_1 & -\omega_1 & & & \vdots \\ 0 & \omega_2 & -\omega_2 & & \vdots \\ \vdots & & \ddots & \ddots & 0 \\ 0 & \cdots & 0 & \omega_N & -\omega_N \end{bmatrix}$$

$$B = \begin{bmatrix} 1 & 0 & \cdots & \cdots & 0 \\ -\alpha & 1 & & & \vdots \\ 0 & -\alpha & 1 & & \vdots \\ \vdots & & \ddots & \ddots & 0 \\ 0 & \cdots & 0 & -\alpha & 1 \end{bmatrix}\begin{bmatrix} C_0 \\ 0 \\ \vdots \\ \vdots \\ 0 \end{bmatrix} \qquad (5.4)$$

$$C = [0 \; \cdots \; 0 \; 1]$$

Time domain fractional order system identification is mainly based on two class of models viz., equation-error models and output-error models.

5.1.1 Equation Error Model

Let us consider, the dynamical system is governed by the following fractional order differential equation

$$
\begin{aligned}
&y(t) + b_1 D^{\beta_1} y(t) + \cdots + b_{m_B} D^{\beta_{m_B}} y(t) \\
&= a_0 D^{\alpha_0} u(t) + a_1 D^{\alpha_1} u(t) + \cdots + a_{m_A} D^{\alpha_{m_A}} u(t)
\end{aligned}
\tag{5.5}
$$

with non-integer differentiation order $\beta_1 < \beta_2 < \cdots < \beta_{m_B}, \alpha_0 < \alpha_1 < \cdots < \alpha_{m_A}$. The equation-error based models are based on linear coefficients. The system to be identified is considered to be initially at rest and is characterized by the parameter vector

$$
\theta = \begin{bmatrix} a_0 & \cdots & a_{m_A} & b_1 & \cdots & b_{m_B} \end{bmatrix}^T
\tag{5.6}
$$

In this method a commensurate order γ is chosen and the differential orders in the model are fixed as a multiple of γ up to the maximum order of the numerator and denominator of the FO transfer function. The maximum numerator and denominator orders are chosen judiciously ($\alpha_{m_A} \leq \beta_{m_B} - \gamma$) so as to ensure a strictly proper transfer-function model.

$$
F(s) = \frac{\sum_{k=0}^{\frac{\alpha_{m_A}}{\gamma}} a_k s^{k\gamma}}{1 + \sum_{j=1}^{\frac{\beta_{m_B}}{\gamma}} b_j s^{j\gamma}}
\tag{5.7}
$$

Let us consider, that the observed input data $u(t)$ and output data $y^*(t)$ are collected at regular samples: $k_0 T_s, (k_0 + 1) T_s, \cdots, (k_0 + K - 1) T_s$.

where $y^*(t) = y(t) + p(t)$ and $p(t)$ is the perturbation signal. Now, the estimation method consists of computing fractional derivatives of sampled input and output data using the Grunwald-Letnikov definition of fractional differentiation. The system output can be expressed as the following regression form:

$$
y(t) = \phi^*(t) \theta
\tag{5.8}
$$

where θ is the parameter vector and the regression vector is given by

$$
\phi^*(t) = \begin{bmatrix} D^{\alpha_0} u(t) & \cdots & D^{\alpha_{m_A}} u(t) & -D^{\beta_1} y^*(t) & \cdots & -D^{\beta_{m_B}} y^*(t) \end{bmatrix}
\tag{5.9}
$$

The estimated parameter vector $\hat{\theta}$ is obtained by minimizing the quadratic norm of estimation error:

$$
J\left(\hat{\theta}\right) = E^T E
\tag{5.10}
$$

where the error vector is composed as:

$$E = [\varepsilon\,(k_0 T_s) \quad \varepsilon\,((k_0 + 1)\,T_s) \quad \cdots \quad \varepsilon\,((k_0 + K - 1)\,T_s)]^T \tag{5.11}$$

and

$$\varepsilon\,(t) = y^*\,(t) - \phi^*\,(t)\,\hat{\theta} \tag{5.12}$$

The classical least-square method can now be used to obtain the minimum of J.

$$\hat{\theta}_{\text{opt}} = (\Phi^{*T}\Phi^*)^{-1}\Phi^{*T}Y^* \tag{5.13}$$

where

$$\Phi^* = \left[\phi^{*T}\,(k_0 T_s) \quad \phi^{*T}\,((k_0 + 1)\,T_s) \quad \cdots \quad \phi^{*T}\,((k_0 + K - 1)\,T_s)\right]^T \tag{5.14}$$

As in the integer order case, fractional differentiation also amplifies noisy signals. Hence, a linear low-pass filter is designed so as to obtain a linear continuous regression of the filtered input $u_f\,(t)$ and output $y_f^*\,(t)$ signals:

$$y_f\,(t) = \phi_f^*\,(t)\,\theta \tag{5.15}$$

where

$$\phi_f^*\,(t) =$$
$$\left[D^{\alpha_0}u_f\,(t) \quad \cdots \quad D^{\alpha_{m_A}}u_f\,(t) \quad -D^{\beta_1}y_f^*\,(t) \quad \cdots \quad -D^{\beta_{m_B}}y_f^*\,(t)\right] \tag{5.16}$$

The filter is generally chosen to be causal, stationary and low-pass. Among the possible filters the linear integral filter is most common:

$$H\,(s) = \left(\frac{1}{s}\right)^L \tag{5.17}$$

where the order L is generally chosen as the highest order of differentiation in the model to be identified i.e.$L = \beta_{m_B}$. In this modification, the system's governing differential equation can be rewritten as the following integral equation:

$$D^{-\beta_{m_B}}y\,(t) + b_1 D^{\beta_1 - \beta_{m_B}}y\,(t) + \cdots + b_{m_B}y\,(t)$$
$$= a_0 D^{\alpha_0 - \beta_{m_B}}u\,(t) + a_1 D^{\alpha_1 - \beta_{m_B}}u\,(t) + \cdots + a_{m_A}D^{\alpha_{m_A} - \beta_{m_B}}u\,(t) \tag{5.18}$$

Here, the negative power of D denotes an integral operator.

This particular method avoids differentiation of noisy signals; however it integrates noise, producing wrong steady-state and low dynamics estimations. To eliminate this problem the following fractional state variable filter can be used:

$$H\,(s) = \frac{A}{\alpha_0 + \alpha_1 s^\gamma + \cdots + \alpha_{N_f - 1}s^{\gamma(N_f - 1)} + s^{\gamma N_f}} \tag{5.19}$$

Here, γN_f is the filter's order. The filter design must meet the following design specifications viz. $N_f > \max\left(\beta_{m_B}, \alpha_{m_A}\right)$ and coefficients $\left\{\alpha_0, \alpha_1, \cdots, \alpha_{N_f-1}\right\}$ must be chosen such that $H(s)$ is stable. A particular choice of the state variable filter can be the fractional Poisson's filter:

$$H(s) = \frac{1}{\left(1 + \left(\frac{s}{\omega_f}\right)^{\gamma}\right)^{N_f}}$$

$$= \frac{\omega_f^{\gamma N_f}}{s^{\gamma N_f} + \binom{N_f}{1}\omega_f^{\gamma} s^{\gamma(N_f-1)} + \cdots + \binom{N_f}{N_f-1}\omega_f^{\gamma(N_f-1)} s^{\gamma} + \omega_f^{\gamma N_f}}$$

$$(5.20)$$

Frequency ω_f is fixed by the user close to the highest corner frequency of the system. The state-vector, composed of filtered input and output is defined as:

$$x_f = \left[D^{(N_f-1)\gamma} z_f(t), \quad D^{(N_f-2)\gamma} z_f(t), \quad \cdots \quad z_f(t)\right] \qquad (5.21)$$

Here, z_f denotes either the input (u_f) or output (y_f). The fractional order state space realization of the filter is given as:

$$D^{\gamma} x_f(t) = A_f x_f(t) + B_f z_f(t) \qquad (5.22)$$

where

$$A_f = \begin{bmatrix} \binom{N_f}{1}\omega_f^{\gamma} & \binom{N_f}{2}\omega_f^{2\gamma} & \cdots & \binom{N_f}{N_f-1}\omega_f^{\gamma}(N_f-1) & \omega_f^{\gamma N_f} \\ -1 & 0 & \cdots & \cdots & 0 \\ 0 & \ddots & 0 & & \vdots \\ \vdots & \ddots & \ddots & \ddots & \vdots \\ 0 & \cdots & 0 & -1 & 0 \end{bmatrix}$$

and

$$B_f = \left[\omega_f^{\gamma N_f} \quad 0 \quad \cdots \quad 0\right].$$

The fractional Poisson filter can be simulated by the well known Grunwald–Letnikov definition of fractional differentiation. The estimated parameter vector can now be obtained by minimizing the quadratic norms of the filtered equation error:

$$J\left(\hat{\theta}\right) = E_f^T E_f \qquad (5.23)$$

where

$$E_f = \left[\varepsilon(k_0 T_s) \quad \varepsilon((k_0+1)T_s) \quad \cdots \quad \varepsilon((k_0+K-1)T_s)\right]^T \qquad (5.24)$$

and

$$\varepsilon(t) = y_f^*(t) - \phi_f^*(t)\,\hat{\theta} \tag{5.25}$$

The solution is now given by the classical least square:

$$\hat{\theta} = \left(\Phi_f^T \Phi_f\right)^{-1} \Phi_f^T Y_f^* \tag{5.26}$$

where $\Phi_f^* = \left[\phi_f^{*T}(k_0 T_s) \quad \phi_f^{*T}((k_0 + 1) T_s) \quad \cdots \quad \phi_f^{*T}((k_0 + K - 1) T_s)\right]$

Similar to the classical case, the above least square estimator is biased in the presence of noisy output. To eliminate the bias the conventional instrumental variable technique can be used:

$$\hat{\theta}_{\text{opt}}^{IV} = \left(\Phi_f^{IV^T} \Phi_f^*\right)^{-1} \Phi_f^{IV^T} Y_f^* \tag{5.27}$$

Here, Φ_f^{IV} is the regression matrix, composed of derivatives of the filtered inputs and derivatives of instrumental variables. The instruments can also be optimized to achieve an accurate estimate.

5.1.2 Output Error Model

This technique allows simultaneous estimation of the model orders and the corresponding parameters. This technique is generally applied in three different forms viz., discrete-time simulation based, continuous-time simulation based and fractional orthogonal function based.

5.1.2.1 Discrete Time Simulation Based Output Error Model

The system is considered to be initially at rest and the parameter vector in this method is formed of the coefficients and orders of the numerator and denominator i.e. $\theta = \left[a_0, \cdots a_{m_A}, b_1, \cdots b_{m_B}, \alpha_0, \cdots \alpha_{m_A}, \beta_1, \cdots \beta_{m_B}\right]$. For a commensurate order system, if the order of the system is very high or the commensurate order γ is very small, the number of parameters to be estimated becomes large and the optimization becomes ill-conditioned. To overcome this problem, the orders α_{m_A} and β_{m_B} can be fixed as multiples of γ. The system is then entirely characterized by the coefficient vector $\theta = \left[a_0, \cdots a_{m_A}, b_1, \cdots b_{m_B}, \gamma\right]$. For identification of stable systems the commensurate order can be constrained to $]0, 2[$.

Considering observed data $u(t)$ and $y^*(t) = y(t) + p(t)$, $p(t)$ being an output white noise, the quadratic norm

$$J\left(\hat{\theta}\right) = \sum_{k=k_0}^{k_0+K-1} \varepsilon^2\left(kT_s, \hat{\theta}\right) \tag{5.28}$$

of output error

$$\varepsilon \left(kT_s, \hat{\theta} \right) = y^* \left(kT_s \right) - \hat{y} \left(kT_s, \hat{\theta} \right) \tag{5.29}$$

is now minimized. Since, the model's output $\hat{y} \left(kT_s, \hat{\theta} \right)$ is nonlinear in $\hat{\theta}$, gradient-based algorithms, such as Marquardt algorithm can be used to estimate $\hat{\theta}$ iteratively using the following equations.

$$\hat{\theta}_{i+1} = \hat{\theta}_i - \left\{ \left[J''_{\theta\theta} + \xi I \right]^{-1} J'_\theta \right\}_{\theta = \hat{\theta}_i} \tag{5.30}$$

where the gradient is

$$J'_\theta = -2 \sum_{k=k_0}^{k_0+K-1} \varepsilon \left(kT_s \right) S \left(kT_s, \hat{\theta} \right) \tag{5.31}$$

The Pseudo–Hessian is

$$J''_{\theta\theta} \approx -2 \sum_{k=k_0}^{k_0+K-1} S \left(kT_s, \hat{\theta} \right) S^T \left(kT_s, \hat{\theta} \right) \tag{5.32}$$

The Output sensitivity function is

$$S \left(kT_s, \hat{\theta} \right) = \frac{\partial \hat{y} \left(kT_s, \hat{\theta} \right)}{\partial \theta} \tag{5.33}$$

and the Marquardt parameter is ξ.

5.1.2.2 Continuous Time Simulation Based Output Error Model

The equation given by (5.1), is a fractional order integrator bounded in the frequency band. The differentiation order is estimated by keeping α fixed and calculating η of Eq. 5.1. When α and η are found, the order γ can be calculated from Eq. 5.2.

Considering the fractional differential system

$$D^\gamma y(t) + a_0 y(t) = b_0 u(t) \tag{5.34}$$

and defining $X(s)$ as

$$X(s) = \frac{U(s)}{s^n + a_0} \tag{5.35}$$

The generalized state space representation of this system is given by

$$D^\gamma y\left(t\right) = -a_0 y\left(t\right) + u\left(t\right)$$
$$y\left(t\right) = b_0 x\left(t\right) \tag{5.36}$$

As shown in Sect. 5.1 the equivalent continuous time state space model of (5.36) can be represented as

$$x_1'\left(t\right) = -a_0 x_{N+1}\left(t\right) + u\left(t\right)$$
$$y\left(t\right) = b_0 x_{N+1}\left(t\right) \tag{5.37}$$

From the state space representation of the continuous time approximation, the following global rational state space representation can be obtained:

$$x_G' = A_G x_G + B_G u$$
$$y = C_G x_G \tag{5.38}$$

where

$$A_G = A + \begin{bmatrix} 0 & \cdots & 0 & -a_0 \\ \vdots & \ddots & & 0 \\ \vdots & & \ddots & \vdots \\ 0 & \cdots & 0 & 0 \end{bmatrix}$$

$$B_G = B, \quad C_G^T = \begin{bmatrix} 0 & \cdots & 0 & b_0 \end{bmatrix}$$

and A and B being defined in Eq. 5.3.

For the fractional integrator bounded in the frequency band$[\omega_A, \omega_B]$, α is fixed and the parameter vector

$$\theta^T = [a_0, \ b_0, \ \eta] \tag{5.39}$$

is estimated.

The objective function is defined by Eqs. 5.28 and 5.29. The coefficients can be calculated by the recursive Marquardt algorithm as in Eq. 5.30. Sensitivity functions can be found by calculating the partial derivatives of (5.1) with respect to each of the parameters of (5.39).

5.1.2.3 Orthogonal Function Based Output Error Model

A complete orthogonal basis in $L_2\left[0, \infty\right[$ (Lebesgue space of square integrable funtions) have been synthesized (Aoun et al. 2007). Laguerre, Kautz and GOB functions are hence extended to fractional differentiation orders. Thus any finite energy stable transfer function $F\left(s\right)$ belonging to the Hardy space $H_2\left(C^+\right)$ can be written as a linear combination of fractional orthogonal functions $G_m\left(s\right)$.

$$F(s) = \sum_{m=1}^{\infty} a_m G_m(s) \tag{5.40}$$

Equation 5.40 is truncated to an order N and $F(s)$ is approximated by

$$F(s) \approx F_M(s) = \sum_{m=1}^{M} a_m G_m(s) \tag{5.41}$$

Let $u(t), y(t), t \in [0, T]$ be the input and output data obtained from a finite energy linear fractional model. The identification method consists of computing the optimal coefficient vector $g = [a_1, a_2, \ldots, a_N]^T$, obtained by the least square minimization of

$$J = \frac{1}{T} \int_0^T (\varepsilon(t))^2 \, dt \tag{5.42}$$

where $\varepsilon(t) = y(t) - \sum_{m=1}^{M} a_m u_{G_m}(t)$
and $y(t)$ and $u_{G_m}(t)$ are respectively the system and the orthogonal network outputs,

$$u_{G_m}(t) = G_m(t) \otimes u(t) \tag{5.43}$$

Setting

$$u_G(t) = \begin{bmatrix} u_{G_1}(t) & u_{G_2}(t) & \cdots u_{G_M}(t) \end{bmatrix} \tag{5.44}$$

the optimum estimation of the Fourier coefficients $\hat{g}$ is given by

$$\hat{g} = \left[\int_0^T u_G(t)^T u_G(t) \, dt \right]^{-1} \int_0^T u_G(t)^T y(t) \, dt \tag{5.45}$$

5.2 Frequency Domain Identification

5.2.1 Levy's Identification Method for Fractional Order Systems

Assuming a linear system is described by a transfer function G having a frequency response $G(j\omega)$, the identification consists of finding out another transfer function of the form

$$\hat{G}\left(s\right) = \frac{b_0 + b_1 s^q + b_2 s^{2q} + \cdots + b_m s^{mq}}{a_0 + a_1 s^q + a_2 s^{2q} + \cdots + a_n s^{nq}} = \frac{\sum_{k=0}^{m} b_k s^{kq}}{\sum_{k=0}^{n} a_k s^{kq}} \tag{5.46}$$

where the orders m and n of the numerator and denominator respectively are user specified and q is the fractional derivative order. Without loss of generality, setting $a_0 = 1$ in (5.46) we have the corresponding frequency response of the identified model as

$$\hat{G}\left(j\omega\right) = \frac{\sum_{k=0}^{m} b_k \left(j\omega\right)^{kq}}{\sum_{k=0}^{n} a_k \left(j\omega\right)^{kq}} = \frac{N\left(j\omega\right)}{D\left(j\omega\right)} = \frac{\alpha\left(\omega\right) + j\beta\left(\omega\right)}{\sigma\left(\omega\right) + j\tau\left(\omega\right)} \tag{5.47}$$

where N and D are complex valued and $\alpha, \beta, \sigma, \tau$ are real valued. From (5.47) we have

$$\alpha\left(\omega\right) = \sum_{k=0}^{m} b_k \mathrm{Re}\left[\left(j\omega\right)^{kq}\right] \tag{5.48}$$

$$\sigma\left(\omega\right) = \sum_{k=0}^{n} a_k \mathrm{Re}\left[\left(j\omega\right)^{kq}\right] = 1 + \sum_{k=1}^{n} a_k \mathrm{Re}\left[\left(j\omega\right)^{kq}\right] \tag{5.49}$$

$$\beta\left(\omega\right) = \sum_{k=0}^{m} b_k \mathrm{Im}\left[\left(j\omega\right)^{kq}\right] \tag{5.50}$$

$$\tau\left(\omega\right) = \sum_{k=0}^{n} a_k \mathrm{Im}\left[\left(j\omega\right)^{kq}\right] = \sum_{k=1}^{n} a_k \mathrm{Im}\left[\left(j\omega\right)^{kq}\right] \tag{5.51}$$

The error between the identified model and the actual system is then given by

$$\varepsilon\left(j\omega\right) = G\left(j\omega\right) - \frac{N\left(j\omega\right)}{D\left(j\omega\right)} \tag{5.52}$$

Since it is difficult to choose parameters in (5.46) such that the error in (5.52) is minimized, Levy's method minimizes the square of the norm of

$$E\left(j\omega\right) := \varepsilon\left(j\omega\right) D\left(j\omega\right) = G\left(j\omega\right) D\left(j\omega\right) - N\left(j\omega\right) \tag{5.53}$$

which gives a set of normal equations having a simpler solution method.

Dropping the frequency argument ω to obtain a simple notation of (5.53) we have

$$
\begin{aligned}
E &= GD - N \\
&= [\mathrm{Re}\,(G) + j\mathrm{Im}\,(G)]\,(\sigma + j\tau) - (\alpha + j\beta) \\
&= [\mathrm{Re}\,(G)\,\sigma - \mathrm{Im}\,(G)\,\tau - \alpha] + j\,[\mathrm{Re}\,(G)\,\tau + \mathrm{Im}\,(G)\,\sigma - \beta]
\end{aligned}
\tag{5.54}
$$

Differentiating Eq. 5.55

$$
|E|^2 = [\mathrm{Re}\,(G)\,\sigma - \mathrm{Im}\,(G)\,\tau - \alpha]^2 + [\mathrm{Re}\,(G)\,\tau + \mathrm{Im}\,(G)\,\sigma - \beta]^2
\tag{5.55}
$$

with respect to one of the coefficients b_k, $k \in \{0, 1, \ldots, m\}$ or a_k, $k \in \{0, 1, \ldots, n\}$, and putting the derivative as zero, we have

$$
\frac{\partial\,|E|^2}{\partial b_k} = 0
\tag{5.56}
$$

i.e.,

$$
[\mathrm{Re}\,(G)\,\sigma - \mathrm{Im}\,(G)\,\tau - \alpha]\,\mathrm{Re}\left[(j\omega)^{kq}\right] + [\mathrm{Re}\,(G)\,\tau + \mathrm{Im}\,(G)\,\sigma - \beta]\,\mathrm{Im}\left[(j\omega)^{kq}\right] = 0
\tag{5.57}
$$

$$
\frac{\partial\,|E|^2}{\partial a_k} = 0
\tag{5.58}
$$

i.e.,

$$
\begin{aligned}
&\sigma\left\{[\mathrm{Im}\,(G)]^2 + [\mathrm{Re}\,(G)]^2\right\}\mathrm{Re}\left[(j\omega)^{kq}\right] \\
&+ \tau\left\{[\mathrm{Im}\,(G)]^2 + [\mathrm{Re}\,(G)]^2\right\}\mathrm{Im}\left[(j\omega)^{kq}\right] \\
&+ \alpha\left\{\mathrm{Im}\,(G)\,\mathrm{Im}\left[(j\omega)^{kq}\right] - \mathrm{Re}\,(G)\,\mathrm{Re}\left[(j\omega)^{kq}\right]\right\} \\
&+ \beta\left\{-\mathrm{Im}\,(G)\,\mathrm{Re}\left[(j\omega)^{kq}\right] - \mathrm{Re}\,(G)\,\mathrm{Im}\left[(j\omega)^{kq}\right]\right\} = 0
\end{aligned}
\tag{5.59}
$$

The $m+1$ equations obtained from (5.57) and n equations obtained from (5.59) form a linear system which can be solved to find the coefficients of (5.46)

$$
\begin{bmatrix} A & B \\ C & D \end{bmatrix}\begin{bmatrix} b \\ a \end{bmatrix} = \begin{bmatrix} e \\ g \end{bmatrix}
\tag{5.60}
$$

$$
\begin{aligned}
A_{l,c} &= -\mathrm{Re}\left[(j\omega)^{lq}\right]\mathrm{Re}\left[(j\omega)^{cq}\right] - \mathrm{Im}\left[(j\omega)^{lq}\right]\mathrm{Im}\left[(j\omega)^{cq}\right] \\
l &= 0, \cdots, m \wedge c = 0, \cdots, m
\end{aligned}
\tag{5.61}
$$

$$
\begin{aligned}
B_{l,c} = {} & \mathrm{Re}\left[(j\omega)^{lq}\right]\mathrm{Re}\left[(j\omega)^{cq}\right]\mathrm{Re}\left[G\left(j\omega\right)\right] + \mathrm{Im}\left[(j\omega)^{lq}\right] \\
& \times \mathrm{Im}\left[(j\omega)^{cq}\right]\mathrm{Im}\left[G\left(j\omega\right)\right] - \mathrm{Re}\left[(j\omega)^{lq}\right]\mathrm{Im}\left[(j\omega)^{cq}\right] \\
& \times \mathrm{Im}\left[G\left(j\omega\right)\right] + \mathrm{Im}\left[(j\omega)^{lq}\right]\mathrm{Im}\left[(j\omega)^{cq}\right]\mathrm{Re}\left[G\left(j\omega\right)\right]
\end{aligned}
\tag{5.62}
$$
$$
l = 0, \cdots, m \wedge c = 1, \cdots, n
$$

$$
\begin{aligned}
C_{l,c} = {} & -\mathrm{Re}\left[(j\omega)^{lq}\right]\mathrm{Re}\left[(j\omega)^{cq}\right]\mathrm{Re}\left[G\left(j\omega\right)\right] + \mathrm{Im}\left[(j\omega)^{lq}\right] \\
& \times \mathrm{Re}\left[(j\omega)^{cq}\right]\mathrm{Im}\left[G\left(j\omega\right)\right] - \mathrm{Re}\left[(j\omega)^{lq}\right]\mathrm{Im}\left[(j\omega)^{cq}\right] \\
& \times \mathrm{Im}\left[G\left(j\omega\right)\right] - \mathrm{Im}\left[(j\omega)^{lq}\right]\mathrm{Im}\left[(j\omega)^{cq}\right]\mathrm{Re}\left[G\left(j\omega\right)\right]
\end{aligned}
\tag{5.63}
$$
$$
l = 1, \cdots, n \wedge c = 0, \cdots, m
$$

$$
\begin{aligned}
D_{l,c} = {} & \left(\{\mathrm{Re}\left[G\left(j\omega\right)\right]\}^2 + \{\mathrm{Im}\left[G\left(j\omega\right)\right]\}^2\right) \times \left\{\mathrm{Re}\left[(j\omega)^{lq}\right]\right. \\
& \left. \times \mathrm{Re}\left[(j\omega)^{cq}\right] + \mathrm{Im}\left[(j\omega)^{lq}\right]\mathrm{Im}\left[(j\omega)^{cq}\right]\right\}
\end{aligned}
\tag{5.64}
$$
$$
l = 1, \cdots, n \wedge c = 1, \cdots, n
$$

$$
e_{l,1} = -\mathrm{Re}\left[(j\omega)^{lq}\right]\mathrm{Re}\left[G\left(j\omega\right)\right] - \mathrm{Im}\left[(j\omega)^{lq}\right]\mathrm{Im}\left[G\left(j\omega\right)\right]
\tag{5.65}
$$
$$
l = 0, \cdots, m
$$

$$
g_{l,1} = -\mathrm{Re}\left[(j\omega)^{lq}\right]\left(\{\mathrm{Re}\left[G\left(j\omega\right)\right]\}^2 + \{\mathrm{Im}\left[G\left(j\omega\right)\right]\}^2\right)
\tag{5.66}
$$
$$
l = 1, \cdots, n
$$

$$
b = \begin{bmatrix} b_0 & \cdots & b_m \end{bmatrix}^T, a = \begin{bmatrix} a_1 & \cdots & a_n \end{bmatrix}^T
\tag{5.67}
$$

(where $\wedge$ represents the Boolean operation of conjunction). For further details please refer to Valério and Sá da Costa (2005); Valério and Costa (2007); Valerio et al. (2008) and the references therein.

5.2.2 Managing Multiple Frequencies

Theoretically speaking, data from one frequency is sufficient to find a model. But in practice due to noise and other measurement inaccuracies, it is desirable to know the frequency response of the plant at more than one frequency to obtain a good identified model. There are two different approaches to deal with data from f frequencies.

The first approach is to sum the systems for each frequency. In this case the matrices A, B, C, D and the vectors e and g in (5.60) is replaced by

$$\tilde{A} = \sum_{p=1}^{f} A_p, \quad \tilde{B} = \sum_{p=1}^{f} B_p, \quad \tilde{C} = \sum_{p=1}^{f} C_p,$$

$$\tilde{D} = \sum_{p=1}^{f} D_p, \quad \tilde{e} = \sum_{p=1}^{f} e_p, \quad \tilde{g} = \sum_{p=1}^{f} g_p.$$

$$(5.68)$$

where A_p, B_p, C_p, D_p, e_p, g_p are given by (5.61)–(5.67) for a particular frequency ω_p. i.e., $A_p := A(\omega_p)$ and others follow similarly.

The second way is to stack several systems to obtain an over-defined system. The pseudo-inverse $([\cdot]^+)$ can be used to obtain a solution to this. Thus Eq. 5.60 becomes

$$\begin{bmatrix} A_1 & B_1 \\ C_1 & D_1 \\ A_2 & B_2 \\ C_2 & D_2 \\ \vdots & \vdots \\ A_f & B_f \\ C_f & D_f \end{bmatrix} \begin{bmatrix} b \\ a \end{bmatrix} = \begin{bmatrix} e_1 \\ g_1 \\ e_2 \\ g_2 \\ \vdots \\ e_f \\ g_f \end{bmatrix} \Rightarrow \begin{bmatrix} b \\ a \end{bmatrix} = \begin{bmatrix} A_1 & B_1 \\ C_1 & D_1 \\ A_2 & B_2 \\ C_2 & D_2 \\ \vdots & \vdots \\ A_f & B_f \\ C_f & D_f \end{bmatrix}^+ \begin{bmatrix} e_1 \\ g_1 \\ e_2 \\ g_2 \\ \vdots \\ e_f \\ g_f \end{bmatrix} \quad (5.69)$$

5.2.3 Adaptation of Levy's Algorithm Using Weights

The identification method can be enhanced using weights for each of the f frequencies. Then Eq. 5.68 can be modified with weights to obtain

$$\tilde{A} = \sum_{p=1}^{f} w_p A_p, \quad \tilde{B} = \sum_{p=1}^{f} w_p B_p, \quad \tilde{C} = \sum_{p=1}^{f} w_p C_p,$$

$$\tilde{D} = \sum_{p=1}^{f} w_p D_p, \quad \tilde{e} = \sum_{p=1}^{f} w_p e_p, \quad \tilde{g} = \sum_{p=1}^{f} w_p g_p.$$

$$(5.70)$$

5.2.3.1 Vinagre's Method

Levy's method has a bias and as such often results in models which have a good fit in the high frequency data, but a poor fit in the low frequency data. Weights that decrease with frequency can be used to balance this. One reasonable value of weights is (Valério and Sá da Costa 2005; Valério and Costa 2007; Valerio et al. 2008)

$$
w_p = \begin{cases} \frac{\omega_2 - \omega_1}{2\omega_1^2} & \text{if } p = 1 \\[2ex] \frac{\omega_{p+1} - \omega_{p-1}}{2\omega_p^2} & \text{if } 1 < p < f \\[2ex] \frac{\omega_f - \omega_{f-1}}{2\omega_f^2} & \text{if } p = f \end{cases} \tag{5.71}
$$

5.2.3.2 Sanathanan's and Koerner's Method

Since Levy's method minimizes E instead of ε, it is possible to find the result of minimizing E through several iterations L, where the function to be minimized is

$$
E_L = \frac{GD - N}{D_{L-1}} \tag{5.72}
$$

In the first iteration $D_1 = 1$ is put to obtain Levy's method. Subsequently D_{L-1} is the denominator found in the previous iteration. This corresponds to a weight

$$
w_p = \frac{1}{\left| D_{L-1} \left(\omega_p \right) \right|^2} \tag{5.73}
$$

If the iterations converge, i.e. $D_{L-1} \approx D_L$ we have $E_L \to \varepsilon$. This method also counterbalances the small effect of low frequency data in the final identified model. More details about this method can be found in Sanathanan and Koerner (1963).

5.3 Other Identification Techniques

Other identification techniques for fractional order systems like improved time and frequency domain methods (Djamah et al. 2008; Valério and Sá da Costa 2010, 2011), subspace methods (Wang et al. 2011), use of bode diagrams (Ghanbari and Haeri 2011), continuous order distribution (Hartley and Lorenzo 2003; Nazarian and Haeri 2010) can also be used depending on the specific requirements of the designer. The continuous order system identification is a new philosophy that considers an order distribution instead of a fixed value of fractional order. This is a frequency domain identification technique. Hartley and Lorenzo (2003) proposed the idea for estaimating all pole continuous order transfer function models and (Nazarian and Haeri 2010) generalized the concept for pole-zero models also.

References

Aoun, M., Malti, R., Levron, F., Oustaloup, A.: Synthesis of fractional Laguerre basis for system approximation. Autom. **43**(9), 1640–1648 (2007)

Djamah, T., Mansouri, R., Djennoune, S., Bettayeb, M.: Optimal low order model identification of fractional dynamic systems. Appl. Math. Comput. **206**(2), 543–554 (2008). doi:10.1016/j.amc.2008.05.109

Gabano, J.D., Poinot, T.: Estimation of thermal parameters using fractional modelling. Signal Process. **91**(4), 938–948 (2011). doi:10.1016/j.sigpro.2010.09.013

Gabano, J.D., Poinot, T.: Fractional modelling and identification of thermal systems. Signal Process. **91**(3), 531–541 (2011). doi:10.1016/j.sigpro.2010.02.005

Gabano, J.D., Poinot, T., Kanoun, H.: Identification of a thermal system using continuous linear parameter-varying fractional modelling. Control Theory & Applications, IET **5**(7), 889–899 (2011)

Ghanbari, M., Haeri, M.: Order and pole locator estimation in fractional order systems using bode diagram. Signal Process. **91**(2), 191–202 (2011). doi:10.1016/j.sigpro.2010.06.021

Hartley, T.T., Lorenzo, C.F.: Fractional-order system identification based on continuous order-distributions. Signal Process. **83**(11), 2287–2300 (2003). doi:10.1016/s0165-1684(03)00182-8

Malti, R., Victor, S., Nicolas, O., Oustaloup, A.: System identification using fractional models: state of the art. ASME Conf. Proc. **2007**(4806X), 295–304 (2007). doi:10.1115/detc2007-35332

Nazarian, P., Haeri, M.: Generalization of order distribution concept use in the fractional order system identification. Signal Process. **90**(7), 2243–2252 (2010). doi:10.1016/j.sigpro.2010.02.008

Sabatier, J., Aoun, M., Oustaloup, A., Grégoire, G., Ragot, F., Roy, P.: Fractional system identification for lead acid battery state of charge estimation. Signal Process. **86**(10), 2654–2657 (2006). doi:10.1016/j.sigpro.2006.02.030

Sanathanan, C., Koerner, J.: Transfer function synthesis as a ratio of two complex polynomials. Autom. Control IEEE Trans. **8**(1), 56–58 (1963)

Valério, D., Costa, J.: Identification of fractional models from frequency data. In: Sabatier et al. (eds.) Advances in fractional calculus, part 4, pp. 229–242. Springer (2007). doi:10.1007/978-1-4020-6042-7_16

Valerio, D., Ortigueira, M.D., da Costa, J.S.: Identifying a transfer function from a frequency response. J. Comput. Nonlinear Dyn. **3**(2), 021207–021207 (2008). doi:10.1115/1.2833906

Valério, D., Sá da Costa, J.: Levy's identification method extended to commensurate fractional order transfer functions. In: Fifth EUROMECH Nonlinear Dynamics Conference 2005, pp. 1357–1366. EUROMECH

Valério, D., Sá da Costa, J.: Finding a fractional model from frequency and time responses. Commun. Nonlinear Sci. Numer. Simul. **15**(4), 911–921 (2010). doi:10.1016/j.cnsns.2009.05.014

Valério, D., Sá da Costa, J.: Identifying digital and fractional transfer functions from a frequency response. Int. J. Control **84**(3), 445–457 (2011). doi:10.1080/00207179.2011.560397

Wang, L., Cheng, P., Wang, Y.: Frequency domain subspace identification of commensurate fractional order input time delay systems. Int. J. Control Autom. Syst. **9**(2), 310–316 (2011)

Chapter 6
Fractional Order Statistical Signal Processing

Abstract The first part of this chapter introduces the different forms of Kalman filter for fractional order systems. The latter part deals with signal processing with fractional lower order moments for α-stable processes. Some methods of parameter estimation of different α-stable processes are reported. The concept of covariation, which is analogous to covariance for Gaussian signal processing, is introduced and its properties and methods of estimation are discussed.

Keywords Fractional Kalman Filter · Extended Fractional Kalman Filter · Fractional lower order moments · Covariation · Symmetric stable distributions

6.1 Fractional Order Kalman Filter

6.1.1 Brief Overview of the Kalman Filter

Filtering is one of the important pillars of signal processing. It is the process by which unwanted noise can be removed from a signal to obtain the original uncorrupted data. It has many practical uses in a wide range of engineering disciplines like communication, control, embedded systems etc. The Kalman filter is an algorithm that can estimate the state variables of a process. From a mathematical viewpoint the Kalman filter algorithm is an optimal state estimator which estimates the states of a linear system from inaccurate and uncertain observations. The Kalman filter is optimal in the sense that it minimizes the Mean Square Error (MSE) of the estimated parameters if the noise follows a Gaussian distribution. If the noise is not Gaussian and only the mean and standard deviation of the noise is known, the Kalman filter is the best linear estimator, though other non linear estimators may work well, depending on specific cases. The term filter is used in the sense that obtaining the best estimate from noisy data is equivalent to filtering out the noise. However, the Kalman filter not only tries to remove the noise from the measurements, but also projects these measurements onto the state estimate. Kalman filter has gained wide popularity in practical applications due to the following:

1. The recursive nature of the algorithm is suitable for real time implementation as the measurement data can be processed as they arrive as opposed to batch processing.
2. The optimality, simple structure and ease of implementatioan of the Kalman filter helps in obtaining good results in practical applications.

6.1.2 The Fractional Kalman Filter

The integer order Kalman filter is extended for state estimation of Fractional order linear state space systems and is known as the Fractional Kalman Filter (FKF— Sierociuk and Dzielinski 2006b). The fractional order Grunwald–Letnikov difference is given as:

$$\Delta^n x_k = \frac{1}{h^n} \sum_{j=0}^{k} (-1)^j \binom{n}{j} x_{k-j} \tag{6.1}$$

where $n \in \mathbb{R}$ is the order of the fractional difference, $\mathbb{R}$ is the set of real numbers, h is the sampling interval, and k is the number of samples for which the derivative is calculated. The factor $\binom{n}{j}$ can be obtained using

$$\binom{n}{j} = \begin{cases} 1 & \text{for } j = 0 \\ \frac{n(n-1)\cdots(n-j+1)}{j!} & \text{for } j > 0 \end{cases} \tag{6.2}$$

Thus the definition allows one to calculate the discrete equivalent of non-integer order differ-integrals. A positive value of n denotes differentiation of that order and a negative value denotes integration. The original function is obtained for $n = 0$. Now a traditional stochastic discrete state space is of the form

$$\begin{aligned} x_{k+1} &= A x_k + B u_k + \omega_k \\ y_k &= C x_k + v_k \end{aligned} \tag{6.3}$$

where x_k is the state vector, u_k is the system input, y_k is the system output, ω_k is the system noise and v_k is the output noise at the time instant k.

This can be represented by

$$\begin{aligned} \Delta^1 x_{k+1} &= A_d x_k + B u_k + \omega_k \\ x_{k+1} &= \Delta^1 x_{k+1} + x_k \\ y_k &= C x_k + v_k \end{aligned} \tag{6.4}$$

where $\Delta^1 x_{k+1} = x_{k+1} - x_k$ is the first order difference of the sample x_k and $A_d = A - I$ (where I is the identity matrix).

Thus the generalized discrete time linear fractional order stochastic system in state-space representation is as follows (Sierociuk and Dzielinski 2006a):

$$\Delta^\gamma x_{k+1} = A_d x_k + B u_k + \omega_k$$

$$x_{k+1} = \Delta^\gamma x_{k+1} - \sum_{j=1}^{k+1} (-1)^j \gamma_j x_{k+1-j} \qquad (6.5)$$

$$y_k = C x_k + v_k$$

where $\gamma_k = diag \left[\begin{pmatrix} n_1 \\ k \end{pmatrix} \quad \cdots \quad \begin{pmatrix} n_N \\ k \end{pmatrix} \right]$ and $\Delta^\gamma x_{k+1} = \begin{bmatrix} \Delta^{n_1} x_{1,k+1} \\ \vdots \\ \Delta^{n_N} x_{N,k+1} \end{bmatrix}$

Here, $n_1, \ldots, n_N$ are the orders of system equations and N is the number of these equations. Estimation results are obtained by minimizing the following cost function in each step:

$$\hat{x}_k = \arg \min_x \left[(\tilde{x}_k - x) \, \tilde{P}_k^{-1} \, (\tilde{x}_k - x)^T + (\tilde{y}_k - Cx) \, \tilde{R}_k^{-1} \, (\tilde{y}_k - Cx)^T \right] \qquad (6.6)$$

where $\tilde{x}_k = E\left[x_k | z_{k-1}^* \right]$ is the state vector prediction at the time instant k, defined as the random variable x_k conditioned on the measurement stream z_{k-1}^*. Additionally, $\hat{x}_k = E\left[x_k | z_k^* \right]$ is the state vector estimate at the time instant k, defined as the random variable x_k conditioned on the measurement stream z_k^*. The measurement stream z_k^* contains the values of the measurement output $y_0, y_1, \ldots, y_N$ and the input signal $u_0, u_1, \ldots, u_N$.

The prediction of the estimation error covariance matrix is defined as:

$$\tilde{P}_k = E\left[(\tilde{x}_k - x_k)(\tilde{x}_k - x_k)^T \right] \qquad (6.7)$$

The covariance matrix of the output noise v_k is defined as $R_k = E\left[v_k v_k^T \right]$ and the covariance matrix of the system noise ω_k is defined as $Q_k = E\left[\omega_k \omega_k^T \right]$. Additionally, $P_k = E\left[(\hat{x}_k - x_k)(\hat{x}_k - x_k)^T \right]$ is the estimation error covariance matrix. All the covariance matrices are assumed to be symmetric.

The FKF for the discrete fractional order stochastic system in the state-space representation introduced by (6.5) is given by the following set of equations:

$$\Delta^\gamma \tilde{x}_{k+1} = A_d \hat{x}_k + B u_k,$$

$$\tilde{x}_{k+1} = \Delta^\gamma \tilde{x}_{k+1} - \sum_{j=1}^{k+1} (-1)^j \gamma_j \hat{x}_{k+1-j},$$

$$\tilde{P}_k = (A_d + \gamma_1) P_{k-1} (A_d + \gamma_1)^T + Q_{k-1} + \sum_{j=2}^{k} \gamma_j P_{k-j} \gamma_j^T, \qquad (6.8)$$

$$\hat{x}_k = \tilde{x}_k + K_k (y_k - C\tilde{x}_k),$$

$$P_k = (I - K_k C) \tilde{P}_k$$

where $K_k = \tilde{P}_k C^T \left(C \tilde{P}_k C^T + R_k \right)^{-1}$ with the initial conditions

$$x_0 \in \mathbb{R}^N, \quad P_0 = E \left[(\tilde{x}_0 - x_0)(\tilde{x}_0 - x_0)^T \right].$$

Here v_k and ω_k are assumed to be independent and with zero mean.

The two simplifying assumptions used in the derivation of the above FKF are as follows:

(a) $E \left[x_{k+1-j}, z_k^* \right] \simeq E \left[x_{k+1-j}, z_{k+1-j}^* \right]$ for $i = 1, \dots, (k+1)$

(b) $E \left[(\hat{x}_l - x_l)(\hat{x}_m - x_m)^T \right]$ are equal to zero when $l \neq m$.

Expression (a) can be interpreted in the sense that the last state vector is updated in each of the filter iteration. Expression (b) implies that no correlation exists amongst past state vectors. For practical implementation of this algorithm in actual hardware the number of elements in the sum in (6.5) has to be limited to a predefined value L. Thus Eq. 6.5 can be expressed as

$$x_{k+1} = \Delta^\gamma x_{k+1} - \sum_{j=1}^{L} (-1)^j \gamma_j x_{k-j+1} \tag{6.9}$$

A small value of L speeds up calculation but affects the accuracy of the estimate, while a large value gives more accuracy at the cost of increased computation. The choice of L depends on sampling time and system time constants and must be appropriately chosen for each application.

6.1.3 The Extended Fractional Kalman Filter for Non-linear Systems

The FKF is only valid for linear stochastic fractional order systems. The FKF can be modified to be applied for non-linear difference equations and is known as the Extended Fractional Kalman Filter (EFKF). The nonlinear discrete stochastic fractional order system in a state-space representation is given by:

$$\Delta^\gamma x_{k+1} = f(x_k, u_k) + \omega_k,$$

$$x_{k+1} = \Delta^\gamma x_{k+1} - \sum_{j=1}^{k+1} (-1)^j \gamma_j x_{k+1-j}, \tag{6.10}$$

$$y_k = h(x_k) + v_k.$$

For this nonlinear stochastic discrete fractional order state space model the Extended Kalman Filter is given by the following:

$$\Delta^\gamma \tilde{x}_{k+1} = f\left(\hat{x}_k, u_k\right),$$

$$\tilde{x}_{k+1} = \Delta^\gamma \tilde{x}_{k+1} - \sum_{j=1}^{k+1} (-1)^j \gamma_j \hat{x}_{k+1-j},$$

$$\tilde{P}_k = (F_{k-1} + \gamma_1)\, P_{k-1}\, (F_{k-1} + \gamma_1)^T + Q_{k-1} + \sum_{j=2}^{k} \gamma_j P_{k-j} \gamma_j^T, \tag{6.11}$$

$$\hat{x}_k = \tilde{x}_k + K_k\left(y_k - h\left(\tilde{x}_k\right)\right),$$

$$P_k = (I - K_k H_k)\, \tilde{P}_k$$

with the initial conditions $x_0 \in \mathbb{R}^N$, $\quad P_0 = E\left[\left(\hat{x}_0 - x_0\right)\left(\hat{x}_0 - x_0\right)^T\right]$, where

$$K_k = \tilde{P}_k H_k^T \left(H_k \tilde{P}_k H_k^T + R_k\right)^{-1},$$

$$F_{k-1} = \left[\frac{\partial f\left(x, u_{k-1}\right)}{\partial x}\right]_{x=\hat{x}_{k-1}}, \quad H_k = \left[\frac{\partial h\left(x\right)}{\partial x}\right]_{x=\tilde{x}_k} \tag{6.12}$$

and the noise sequences v_k and ω_k are assumed to be independent and zero mean.

The EFKF has found applications in practical estimation problems as in Romanovas et al. (2010) and in chaotic secure communication schemes as in Kiani-B et al. (2009).

6.1.4 Improved Fractional Kalman Filter

The assumptions of the FKF as stated in expressions (a) and (b) of the earlier section make the filter suboptimal. The infinite dimensional form of the discrete fractional state space considering all its previous states can be used to overcome these limitations and obtain an optimal solution of the Kalman filter. But due to the infinite dimensional form, numerical computation is quite difficult. Thus the improved Fractional Kalman filter (ExFKF) (Sierociuk et al. 2011) has been proposed which assumes that only m number of past state vectors are used in the estimation process. This reduces the computational load in comparison to the infinite dimensional case and is an improvement over the FKF.Next, the linear fractional order state space is defined in such a manner that it comprises of m state vectors from time k to $k-m+1$.

$$\mathbb{X}_{k+1} = \mathbb{A}_m \mathbb{X}_k + \mathbb{B}_m u_k + \mathbb{I}\omega_k - \mathbb{I} \sum_{j=m+1}^{k+1} (-1)^j \gamma_j x_{k+1-j} \tag{6.13}$$

$$y_k = \mathbb{C}_m \mathbb{X}_k$$

where

$$\mathbb{X}_k = \begin{bmatrix} x_k \\ x_{k-1} \\ \vdots \\ x_{k-m+1} \end{bmatrix}, \quad \mathbb{I} = \begin{bmatrix} I \\ 0 \\ \vdots \\ 0 \end{bmatrix},$$

$$\mathbb{A}_m = \begin{bmatrix} (A + \gamma_1) & -(-1)^2 \gamma_2 & \cdots & -(-1)^m \gamma_m \\ I & \cdots & 0 & 0 \\ \vdots & \vdots & \vdots & \vdots \\ 0 & \cdots & I & 0 \end{bmatrix}, \quad (6.14)$$

$$\mathbb{B}_m = \begin{bmatrix} B \\ 0 \\ \vdots \\ 0 \end{bmatrix}, \quad \mathbb{C}_m = [C \ 0 \ \cdots \ 0]$$

and $I \in \mathbb{R}^{N \times N}$ is the identity matrix.

The ExFKF for the m-finite form of the discrete fractional order system is as follows:

$$\widetilde{\mathbb{X}}_{k+1} = \mathbb{A}_m \widehat{\mathbb{X}}_k + \mathbb{B}_m u_k - \mathbb{I} \sum_{j=m+1}^{k+1} (-1)^j \gamma_j \hat{x}_{k+1-j}$$

$$\widetilde{\mathbb{P}}_k = \mathbb{A}_m \mathbb{P}_{k-1} \mathbb{A}_m^T + \mathbb{Q}_{k-1} + \sum_{j=m+1}^{k} \mathbb{I} \gamma_j P_{k-j} \gamma_j^T \mathbb{I}^T \quad (6.15)$$

$$\widehat{\mathbb{X}}_k = \widetilde{\mathbb{X}}_k + \mathbb{K}_k \left(y_k - \mathbb{C}_m \widetilde{\mathbb{X}}_k \right)$$

$$\mathbb{P}_k = (I - \mathbb{K}_k \mathbb{C}_m) \widetilde{\mathbb{P}}_k$$

where

$$\mathbb{K}_k = \widetilde{\mathbb{P}}_k \mathbb{C}_m^T \left(\mathbb{C}_m \widetilde{\mathbb{P}}_k \mathbb{C}_m^T + R_k \right)^{-1}$$

$$\mathbb{I} = \begin{bmatrix} I \\ 0 \\ \vdots \\ 0 \end{bmatrix}, \quad \mathbb{Q} = \mathbb{I} Q_k \mathbb{I}^T$$

$$(6.16)$$

$$\mathbb{P}_k = \begin{bmatrix} P_{k,k} & P_{k,k-1} & \cdots & P_{k,k-m} \\ P_{k-1,k} & P_{k-1,k-1} & \cdots & P_{k-1,k-m} \\ \vdots & \vdots & \vdots & \vdots \\ P_{k-m,k} & P_{k-m,k-1} & \cdots & P_{k-m,k-m} \end{bmatrix}$$

with initial conditions

$$\widehat{\mathbb{X}}_0 \in \mathbb{R}^{mN}, \quad \mathbb{P}_0 = E\left[\left(\widehat{\mathbb{X}}_0 - \mathbb{X}_0\right)\left(\widehat{\mathbb{X}}_0 - \mathbb{X}_0\right)^T\right]$$

and noises v_k, ω_k are assumed to be independent with zero expected value and matrices $\mathbb{P}_k$, R_k are positive-defined.

The ExFKF has been successfully applied to an estimation problem over lossy networks in Sierociuk et al. (2011).

6.2 Fractional Lower Order Moments and Their Applications to Signal Processing

6.2.1 Importance of Non-Gaussian Statistical Signal Processing

Traditional statistical signal processing techniques are mostly based on the assumption of the ideal Gaussian model of signal or noise. In many applications this is logical and can be explained from the Central Limit Theorem. However, many applications in signal processing is inherently non-Gaussian and as such systems designed with Gaussian assumptions exhibit severe performance deterioration in such non-Gaussian environments (Tsakalides and Nikias 1998; Tsihrintzis et al. 1996; Arikan et al. 1994; Bodenschatz and Nikias 1997; Tsihrintzis and Nikias 1996; Arce 2005; Zha and Qiu 2007).

Stable distributions as discussed in Chap. 3 are a class of important non-Gaussian models which are a flexible modeling tool and are able to describe a wide variety of modeling phenomena. The tails of the density function of the non-Gaussian stable distributions are heavier than their Gaussian counterpart and are suitable in modeling signals having bursty or impulsive nature. The characteristic exponent $\alpha \in (0, 2)$ of the stable distributions is a measure of the heaviness of the tails. Smaller values of α near 0 implies severe impulsiveness. As $\alpha \to 2$ a more Gaussian type behavior is observed and for $\alpha = 2$ the Gaussian distribution itself is obtained.

However, the stable distributions do not enjoy wide popularity in the signal processing community due to two reasons. Firstly, except for few values of α, the stable distributions do not have explicit expressions for their density or distribution functions and their computation is heavily reliant on infinite series expansions or numerical integration of the Fourier inversion formula based on the characteristic function. But the simplicity of the characteristic function partially compensates this unavailability of closed form expression and many standard statistical procedures like Maximum Likelihood Estimation (MLE) and hypothesis testing can be implemented using these characteristic functions. Secondly, for non-Gaussian stable distribution whose characteristic exponent is α, only moments of order less than α are finite. This implies that variance or the second order moment of these stable distributions with $\alpha < 2$ does not exist and variance cannot be used as a measure of dispersion. Also many other standard signal processing techniques like spectral analysis and least square based methods which are based on the assumption of finite variance will not be appropriate in this case. Especially least square techniques give poor performance if the signal contains outliers (which is inevitable in non-Gaussian environment) mainly due to their non-robustness in the presence of outliers. Thus, other statistical techniques based on Fractional Lower Order Moments (FLOM) (Shao and Nikias 1993) have to be used in these cases which makes non-Gaussian stable signal processing harder and more complicated than their traditional Gaussian counterpart.

Since the variance of a non-Gaussian stable distribution is not finite, the dispersion which is analogous to the variance is used, as a measure of the variability of the stable random variables. The minimum dispersion criterion which is a generalization of the minimum mean squared error (MMSE) criterion is a measure of optimality in stable signal processing. The average magnitude of the estimation error and the probability of the large estimation errors are minimized by minimizing the error dispersion.

FLOM of estimation errors is a measure of the L_p (for $p < \alpha \le 2$) distance between an estimate and its true value. Minimizing the dispersion is actually equivalent to the minimization of the FLOM of the estimation errors. FLOM based stable signal processing brings in non-linearity even in simple linear estimation problems since they have to be solved in the Banach or metric spaces as opposed to the Hilbert space in the Gaussian case. Higher moments have been extended to the fractional and complex cases and analysis has been done in Nigmatullin (2006).

6.2.2 Characteristic Functions, FLOM and Covariations

6.2.2.1 SαS and Their Characteristic Function

The characteristic function of a $S\alpha S$ process of a real random variable X is of the form

$$\varphi(t) = \exp\left\{ jat - \gamma \left|t\right|^{\alpha} \right\} \tag{6.17}$$

where $0 < \alpha \le 2$ is the characteristic exponent, $\gamma > 0$ is the dispersion (same as σ as introduced in Chap. 3) and $-\infty < a < \infty$ is the location parameter (same as μ as introduced in Chap. 3).

The real random variables $X_1, \ldots, X_n$ are jointly $S\alpha S$ if their joint characteristic function is of the form:

$$\varphi(t) = \exp\left\{ jt^T a - \int_S \left| t^T s \right|^{\alpha} \mu_s(ds) \right\} \tag{6.18}$$

where the spectral measure $\mu_s(\cdot)$ is symmetric, $\mu_s(A) = \mu_s(-A)$ for any measurable set A on the unit sphere S. t, a, s are n-dimensional real vectors.

For $1 < \alpha \le 2$, $X_1, \ldots, X_n$ are jointly $S\alpha S$ if and only if all the linear combinations $a_1 X_1 + a_2 X_2 + \cdots + a_n X_n$ are $S\alpha S$.

6.2.2.2 Fractional Lower Order Moments

The second order moments of all $S\alpha S$ random variable with $0 < \alpha < 2$ do not exist. However, all moments less than order α exist and are known as FLOM. The FLOM of a $S\alpha S$ random variable can be expressed in terms of the dispersion γ and characteristic exponent α, as follows:

$$E\,|X|^p = C\,(p,\alpha)\,\gamma^{\frac{p}{\alpha}} \tag{6.19}$$

for $0 < p < \alpha$ where $C\,(p,\alpha) = \frac{2^{p+1}\Gamma\left(\frac{p+1}{2}\right)\Gamma(-p/\alpha)}{\alpha\sqrt{\pi}\Gamma(-p/2)}$, Γ is the gamma function and X is an $S\alpha S$ random variable with zero location parameter and dispersion γ. The norm of X can be defined as:

$$\|X\|_\alpha = \begin{cases} \gamma^{\frac{1}{\alpha}} & \text{for } 1 \le \alpha \le 2 \\ \gamma & \text{for } 0 < \alpha < 1 \end{cases} \tag{6.20}$$

Hence, the norm $\|X\|_\alpha$ is actually a scaled version of the dispersion and quantifies the distribution of X through the characteristic function

$$\varphi\,(t) = \begin{cases} \exp\left\{-\|X\|_\alpha^\alpha\,|t|^\alpha\right\} & \text{for } 1 \le \alpha \le 2 \\ \exp\left\{-\|X\|_\alpha\,|t|^\alpha\right\} & \text{for } 0 < \alpha < 1 \end{cases} \tag{6.21}$$

For two jointly $S\alpha S$ random variables X and Y, the distance between them is defined as:

$$d_\alpha\,(X, Y) = \|X - Y\|_\alpha \tag{6.22}$$

From Eqs. 6.19, 6.20, 6.22 the distance $d_\alpha\,(X, Y)$ can be expressed as

$$d_\alpha\,(X, Y) = \begin{cases} \left(E\,|X - Y|^p / C\,(p,\alpha)\right)^{1/p} & \text{for } 0 < p < \alpha,\ 1 \le \alpha \le 2 \\ \left(E\,|X - Y|^p / C\,(p,\alpha)\right)^{\alpha/p} & \text{for } 0 < p < \alpha,\ 0 < \alpha < 1 \end{cases} \tag{6.23}$$

Thus the distance $d_\alpha\,(X, Y)$ is a measure of the pth-order moments $(0 < p < \alpha)$ of the difference of the two $S\alpha S$ random variables X and Y. For Gaussian distribution $(\alpha = 2)$, d_α is half of the variance of the difference. Also all lower order moments of a $S\alpha S$ random variable are equivalent in the sense that the pth and qth order moments differ by a constant factor irrespective of the $S\alpha S$ random variable for all $0 < p,\ q < \alpha$.

6.2.2.3 Covariations and Their Properties

Traditional signal processing techniques like prediction, filtering and smoothing are based on the concept of covariance. However since $S\alpha S$ random variables do not have a finite variance, hence the covariance do not exist. An equivalent quantity known as covariation has been defined to circumvent this issue.

For two jointly $S\alpha S$ random variables X and Y, with $1 < \alpha \le 2$, zero location parameters and dispersions γ_x and γ_y, respectively, the covariation of X with Y is defined as:

$$[X, Y]_\alpha = \frac{E\left(XY^{\langle p-1 \rangle}\right)}{E\left(|Y|^p\right)}\gamma_y \tag{6.24}$$

where for any real number z and $a \ge 0$, the following convention is used $z^{\langle a \rangle} = |z|^a$ sign (z) and sign $(\cdot)$ represents the Signum function.

As long as $1 \le p < \alpha$, the above definition can be proved to be independent of the value of p.

Also it follows that

$$[X, X]_\alpha = \gamma_x = \|X\|_\alpha^\alpha \tag{6.25}$$

$$[Y, Y]_\alpha = \gamma_y = \|Y\|_\alpha^\alpha \tag{6.26}$$

The covariation coefficient of X with respect to Y can be defined as

$$\lambda_{X,Y} = \frac{[X, Y]_\alpha}{[Y, Y]_\alpha} = \frac{E\left(XY^{\langle p-1 \rangle}\right)}{E\left(|Y|^p\right)} \tag{6.27}$$

for any $1 \le p < \alpha$.

A few important properties of covariation are as follows:

1. If X and Y are independent and jointly $S\alpha S$ then $[X, Y]_\alpha = 0$, while the converse is not true in general.
2. When $\alpha = 2$, i.e., when X, Y are jointly Gaussian with zero mean, the covariation of X with Y is half of the covariance of X and Y.
3. The Cauchy–Schwartz inequality for any jointly $S\alpha S$ random variables X and Y hold as $|[X, Y]_\alpha| \le \|X\|_\alpha \|Y\|_\alpha^{\langle \alpha-1 \rangle}$.
4. The covariation $[X, Y]_\alpha$ is linear in X, i.e. if X_1, X_2, Y are jointly $S\alpha S$ then $[aX1 + bX_2, Y]_\alpha = a[X_1, Y]_\alpha + b[X_2, Y]_\alpha$, holds for any real constants a and b.
5. In general $[X, Y]_\alpha$ is not linear with respect to the second variable Y. But the pseudo-linearity property with respect to Y holds, i.e., if Y_1, Y_2 are independent and if X, Y_1, Y_2 are jointly $S\alpha S$ then $[X, aY_1 + bY_2]_\alpha = a^{\langle \alpha-1 \rangle}[X, Y_1]_\alpha + b^{\langle \alpha-1 \rangle}[X, Y_2]_\alpha$, holds for any real constants a and b.

6.2.3 Parameter Estimation for Symmetric Stable Distributions

Symmetric stable distributions are parameterized by three variables, viz., the characteristic exponent α, dispersion γ and location parameter a. A problem of practical interest is to estimate these parameters from a measured sequence of a $S\alpha S$ random variable. Since a closed form density function for most of the $S\alpha S$ random variables do not exist, most methods of statistics like the MLE, etc. fail in this case. Suboptimal numerical methods are generally used in practice to deal with this problem.

6.2.3.1 Method of Maximum Likelihood

The standard symmetric stable density function is given by Zolotarev (1986):

$$f_\alpha(x) = \frac{\alpha}{|1-\alpha|\pi} x^{1/(\alpha-1)} \int_0^{\pi/2} v(\theta) e^{-x^{\alpha/(\alpha-1)}v(\theta)} d\theta \quad \text{for } \alpha \neq 1, \; x > 0 \quad (6.28)$$

where $v(\theta) = \frac{1}{(\sin\alpha\theta)^{\alpha/(\alpha-1)}} \cos[(\alpha-1)\theta](\cos\theta)^{1/(\alpha-1)}$

Moreover,

$$f_1(x) = \frac{1}{\pi(1+x^2)} \quad (6.29)$$

$$f_\alpha(0) = \frac{1}{\pi}\Gamma((\alpha+1)/\alpha) \quad (6.30)$$

$$f_2(x) = \frac{1}{2\sqrt{\pi}}e^{-x^2/4} \quad (6.31)$$

Thus the three parameters α, a, c can be estimated from the measurements $x_1, x_2, \ldots, x_N$ by maximizing the log likelihood function

$$\sum_{i=1}^{N} \log[f_\alpha(z_i)] = n\log\alpha - n\log(\alpha-1)\pi + \sum_{i=1}^{N}(\log z_i)/(\alpha-1)$$

$$+ \sum_{i=1}^{N} \log \int_0^{\pi/2} v(\theta) e^{-z_i^{\alpha/(\alpha-1)}v(\theta)} d\theta \quad (6.32)$$

where $z_i = |x_i - a|/c$.

Further details can be found in Brorsen and Yang (1990) and Georgiou and Kyriakakis (2006).

6.2.3.2 Method of Sample Characteristic Function

The sample characteristic function is defined as

$$\widehat{\varphi}(t) = \frac{1}{N}\sum_{k=1}^{N} \exp(jtx_k) \quad (6.33)$$

where N is the sample size and $x_1, \ldots, x_N$ are the observations. It is a consistent estimator of the true characteristic function that uniquely determines the density function. The regression method of (Koutrouvelis 1980, 1981) is based on the following relations between the characteristic function of a $S\alpha S$ distribution and its parameters

$$\log \left(- \log |\varphi(t)|^2 \right) = \log \left(2c^\alpha \right) + \alpha \log |t| \tag{6.34}$$

and

$$\frac{\operatorname{Im} \varphi(t)}{\operatorname{Re} \varphi(t)} = \tan at \tag{6.35}$$

From (6.34) the parameters α and c can be estimated from linear regression as

$$y_k = \mu + \alpha w_k + \varepsilon_k, \quad k = 1, 2, \ldots, K \tag{6.36}$$

where $y_k = \log \left(- \log |\widehat{\varphi}(t_k)|^2 \right)$, $\mu = \log \left(2c^\alpha \right)$, $w_k = \log |t_k|$
$\varepsilon_k, k = 1, \ldots, K$ represent error terms which are assumed to be independent and identically distributed with zero mean. $t_1, \ldots, t_K$ denote an appropriate set of real numbers.

Similarly, the location parameter a can be estimated by linear regression as:

$$z_k = au_k + \varepsilon_k, \quad l = 1, 2, \ldots, L \tag{6.37}$$

where $z_k = \operatorname{Arc} \tan \left(\operatorname{Im} \left(\widehat{\varphi}(u_k) \right) / \operatorname{Re} \left(\widehat{\varphi}(u_k) \right) \right)$ and $u_1, \ldots, u_L$ are an appropriate set of real numbers. The error terms ε_k are again assumed to be independent and identically distributed with zero mean. The whole process is iterated and termination occurs when some pre-specified convergence criteria are met.

6.2.4 Estimation of Covariation

The covariation and the covariation-coefficient play an important role in signal processing with $S\alpha S$ random variables. Thus unbiased and efficient estimators for these are of great significance. In most cases, the covariation-coefficient λ_{XY} is generally required and the value of the covariation $[X, Y]_\alpha$ itself is not necessary. The following methods can be adopted for estimation of the covariations.

6.2.4.1 Fractional Lower Order Moment Estimator

For the independent observations $(X_1, Y_1), \ldots, (X_n, Y_n)$, the FLOM estimator is given by:

$$\widehat{\lambda}_{\text{FLOM}} = \frac{\sum_{i=1}^{N} X_i \, |Y_i|^{p-1} \, \text{sign} \, (Y_i)}{\sum_{i=1}^{N} |Y_i|^p} \tag{6.38}$$

for some $1 \leq p < \alpha$. For computational efficiency, $p = 1$ gives

$$\widehat{\lambda}_{\text{FLOM}} = \frac{\sum_{i=1}^{N} X_i \, \text{sign} \, (Y_i)}{\sum_{i=1}^{N} |Y_i|} \tag{6.39}$$

6.2.4.2 Screened Ratio Estimator

A consistent estimator for covariation coefficients was proposed by (Kanter and Steiger 1974). Assuming X and Y are random variables satisfying $E \, (|X|) < \infty$ and $E \, (X|Y) = \lambda Y$ for some constant λ. For $0 < c_1, c_2 \leq \infty$, defining a new random variable χ_Y from Y we have

$$\chi_Y = \begin{cases} 1 & \text{if } c_1 < |Y| < c_2 \\ 0 & \text{otherwise} \end{cases} \tag{6.40}$$

Then,

$$E\left[\left(XY^{-1}\chi_Y\right)/P \, (c_1 < |Y| < c_2)\right] = \lambda \tag{6.41}$$

For independent observations $(X_1, Y_1), \ldots, (X_N, Y_N)$, the following estimator for λ is strongly consistent

$$\widehat{\lambda}_{SR} = \frac{\sum_{i=1}^{N} \left(X_i Y_i^{-1} \chi_{Y_i}\right)}{\sum_{i=1}^{N} \chi_{Y_i}} \tag{6.42}$$

$\widehat{\lambda}_{SR}$ converges to λ almost surely as $N \to \infty$ where X_i, Y_i are independent copies of X and Y respectively. $\widehat{\lambda}_{SR}$ in (6.42) is called the screened ratio (SR) estimator of λ. The constants c_1, c_2 are arbitrary and generally c_2 is taken to be infinite.

References

Arce, G.R.: Nonlinear Signal Processing: A Statistical Approach. Wiley, New York (2005)

Arikan, O., Enis Cetin, A., Erzin, E.: Adaptive filtering for non-Gaussian stable processes. IEEE Sig. Proc. Lett. 1(11), 163–165 (1994)

Bodenschatz, J.S., Nikias, C.L.: Symmetric alpha-stable filter theory. IEEE Trans. Sig. Proc. 45(9), 2301–2306 (1997)

Brorsen, B.W., Yang, S.R.: Maximum likelihood estimates of symmetric stable distribution parameters. Commun. Stat. Simul. Comput. 19(4), 1459–1464 (1990)

Georgiou, P.G., Kyriakakis, C.: Maximum likelihood parameter estimation under impulsive conditions, a sub-Gaussian signal approach. Sig. Proc. **86**(10), 3061–3075 (2006). doi:10.1016/j.sigpro.2006.01.007

Kanter, M., Steiger, W.: Regression and autoregression with infinite variance. Adv. Appl. Prob. **6**(4), 768–783 (1974)

Kiani-B, A., Fallahi, K., Pariz, N., Leung, H.: A chaotic secure communication scheme using fractional chaotic systems based on an extended fractional Kalman filter. Commun. Nonlinear Sci. Numer. Simul. **14**(3), 863–879 (2009). doi:10.1016/j.cnsns.2007.11.011

Koutrouvelis, I.A.: Regression-type estimation of the parameters of stable laws. J. Am. Stat. Assoc. **75**(372), 918–928 (1980)

Koutrouvelis, I.A.: An iterative procedure for the estimation of the parameters of stable laws. Commun. Stat. Simul. Comput. **10**(1), 17–28 (1981)

Nigmatullin, R.R.: The statistics of the fractional moments: Is there any chance to "read quantitatively" any randomness?. Sig. Proc. **86**(10), 2529–2547 (2006). doi:10.1016/j.sigpro.2006.02.003

Romanovas, M., Klingbeil, L., Traechtler, M., Manoli, Y.: Fractional model based Kalman filters for angular rate estimation in vestibular systems. In: 10th IEEE International Conference on Signal Processing (ICSP), pp. 1–4, 24–28 Oct 2010

Shao, M., Nikias, C.L.: Signal processing with fractional lower order moments: stable processes and their applications. IEEE Proc. **81**(7), 986–1010 (1993)

Sierociuk, D., Dzielinski, A.: Estimation and control of discrete fractional order states-space systems. In: Romaniuk, R.S. (ed.) SPIE, 2006a, ISBN: 61593N, 61510

Sierociuk, D., Dzielinski, A.: Fractional Kalman filter algorithm for the states, parameters and order of fractional system estimation. Int. J. Appl. Math. Comput. Sci. **16**(1), 129 (2006)

Sierociuk, D., Tejado, I., Vinagre, B.M.: Improved fractional Kalman filter and its application to estimation over lossy networks. Sig. Proc. **91**(3), 542–552 (2011). doi:10.1016/j.sigpro.2010.03.014

Tsakalides, P., Nikias, C.L.: Deviation from normality in statistical signal processing: parameter estimation with alpha-stable distributions. In: Alder, R.J., et al. (eds.) A Practical Guide to Heavy Tails: Statistical Techniques and Applications, pp. 379–404. Birkhauser, Boston (1998). ISBN:0-8176-3951-9

Tsihrintzis, G.A., Nikias, C.L.: Alpha-stable impulsive interference: canonical statistical models and design and analysis of maximum likelihood and moment-based signal detection algorithms. In: Cornelius, T.L. (ed.) Control and Dynamic Systems, vol. 78, pp. 341–388. Academic Press, New York (1996)

Tsihrintzis, G.A., Shao, M., Nikias, C.L.: Recent results in applications and processing of [alpha]-stable-distributed time series. J. Franklin Inst. **333**(4), 467–497 (1996). doi:10.1016/0016-0032(96)00027-0

Zha, D., Qiu, T.: Direction finding in non-Gaussian impulsive noise environments. Dig. Sig. Proc. **17**(2), 451–465 (2007). doi:10.1016/j.dsp.2005.11.006

Zolotarev, V.M.: One-Dimensional Stable Distributions. American Mathematical Society, Providence (1986)

Chapter 7
MATLAB Based Simulation Tools

Abstract This chapter mentions several MATLAB based toolboxes and codes which can be used to simulate fractional order signals and systems in general. Most of the toolboxes are in public domain and are freely accessible by any user. The reader can use these tools for simulation right away to get a feel of fractional order signal processing. These tools can also be the building blocks for simulating more complex systems during research.

Keywords MATLAB · Simulink · Ninteger

7.1 Numerical Tools for Fractional Calculus

Numerical solution of the differ-integration can be done by different methods. Numerical solution of fractional order differential or integral equation can be cast as a closed form solution with Mittag-Leffler function and its variants. Podlubny's MATLAB code using two parameter Mitag-Leffler function (Podlubny 2009) has been generalized for four parameter Mittag-Leffler function and the MATLAB code is available in Monje et al. (2010), Sect. 13.1 using *ml_func()*.

Finite difference approach of fractional differ-integrals which is known as the Grunwald–Letnikov definition can be numerically evaluated using the function *glfdiff()* in MATLAB as outlined in Monje et al. (2010), Sect. 13.1 and Xue and Chen (2009). Symbolic computation of fractional differ-integration with Caputo's formula has also been given in Monje et al. (2010), Sect. 13.1. MATLAB based closed form solution of *n*-term linear fractional differential or integral equations can be evaluated using the function *fode_sol()*, by Grunwald–Letnikov definition.

S. Das and I. Pan, *Fractional Order Signal Processing*,
SpringerBriefs in Applied Sciences and Technology,
DOI: 10.1007/978-3-642-23117-9_7, © The Author(s) 2012

7.2 Simulation of FO LTI Systems

Fractional LTI systems can be simulated using the MATLAB based commercial software package CRONE toolbox (Oustaloup et al. 2000). Valerio has made a MATLAB based open source toolbox for simulating fractional order LTI systems known as Ninteger toolbox (Valerio 2005). Ninteger toolbox mainly covers continuous and discrete time realization of FO systems in MATLAB and Simulink, fractional order system identification by frequency domain technique, fractional PID controller design, calculating two-norm and infinity-norm of FO systems, etc. Simulink blocks for simulating fractional differ-integrals and fractional PID controllers are available in Ninteger toolbox. The blocks can be customized with chosen order of approximation, frequency range of fitting, discretization formula and expansion technique. Xue, Chen and Atherton in their book (Xue et al. 2007) has illustrated an object oriented method for simulating fractional order linear systems in their FOTF toolbox which is similar to MATLAB's control system toolbox. This toolbox has evolved over time with added functionalities and can be found at Yang Quan et al. (2009); Xue and Chen (2009); Monje et al. (2010). With the function overloading capability of MATLAB, transfer function simulation, series and parallel connection, feedback control, time and frequency response, evaluation of system norms, adding time delay in the model, etc. can be easily done for FO systems as well, with the same commands of MATLAB's control system toolbox.

7.3 Rational Approximation of FO Differ-integrators

FIR and IIR realization of FO differ-integrators by direct discretization using numerical techniques is discussed in Monje et al. (2010), Chap. 13. Indirect discretization from continuous time filter approximations like Oustaloup's, modified Oustaloup's can also be found in the same chapter using the functions *ousta_fod()* and *new_fod()* (Xue and Chen 2009).

Numerical issues regarding the effect of choosing different s to z plane mapping formula (generating function) as discussed in Chap. 2 can also be found in Monje et al. (2010), Sect. 12.2 and the references therein. MATLAB and Maple based codes are available for CFE based digital realization in Chen and Vinagre (2003); Chen et al. (2004). MATLAB codes for step and impulse response invariant discretization of FO differ-integrators are discussed in Monje et al. (2010), Sect. 12.2.

H_2 norm based sub-optimal model reduction for fractional order transfer function models are given in Monje et al. (2010), Sect. 12.4 and Xue et al. (2007) using the function *opt_app()*. Impulse response invariant discretization of distributed order and fractional second order filter has been developed by Sheng (2010).

7.4 Fractional Fourier and Wavelet Transform

Bultheel and Martínez Sulbaran (2004) have implemented different FrFT, discrete FrFT and sine and cosine transforms in MATLAB. Digital watermarking of images using fractional Fourier transforms have been implemented as MATLAB codes in Bultheel (2007).

The fractional Spline Wavelet Software for MATLAB has been developed by Blu and Unser (1999). This software enables Wavelet based frequency domain analysis and synthesis of low-pass and high-pass filters with various customizations, frequency domain computation of fractional Spline autocorrelation function. FFT based computation of Wavelet and inverse Wavelet transform can also be done with this toolbox.

7.5 Fractional Kalman Filter

Sierociuk has developed a discrete time fractional order stochastic state space system in Simulink using s-function in his Fractional State-Space Toolkit (FSST) (Sierociuk 2005). The toolbox also enables simulation of fractional order state estimators with disturbance, popularly known as Fractional Kalman Filter (FKF).

7.6 System Identification

MATLAB based frequency domain FO system identification can be done with Ninteger toolbox functions (Valerio 2005) and time domain FO system identification with CRONE toolbox functions (Oustaloup et al. 2000).

7.7 Hurst Parameter Estimators

Hurst parameter which is a measure of self-similarity of stochastic variables can be estimated by different estimators like R/S analysis, absolute value method, aggregated variance method, periodogram method, modified periodogram method, difference variance method, Higuchi's method, residuals of regression method, etc. Chen (2008) has developed MATLAB codes for estimating Hurst parameter with the above-mentioned methods.

Other Hurst parameter estimation techniques may include Wavelet estimator and Whittle estimator whose MATLAB implementation can be found at Veitch (2007) and Taqqu. A Java based self-similarity estimator known as Local Analysis of Self-Similarity (LASS) can also be found in Stoev et al. (2006).

7.8 Miscellaneous

The mGn is the output of a variable order integrator driven by white noise. The
MATLAB based simulation tools for variable order integrator has been developed
by Hongguang (2010). ARFIMA code is available at Fatichi (2009). Linear fractional
stable motion and FARIMA time series can be generated using FFT based method
as in Stoev and Taqqu (2004).

Stable distributions can be simulated using software developed by Nolan (2000).
It is possible to calculate stable densities, cumulative distribution functions and quan-
tiles using this software. It also has stable random number generation and maximum
likelihood estimation of stable parameters using a fast three-dimensional cubic Spline
interpolation of stable densities. Software for calculating multivariate stable densities can also be found in Nolan. MATLAB code for simulating Levy and α-stable
distributions can be found in Veillette (2009). A detailed review on other available
MATLAB, Mathematica and different software packages for simulating fractional
calculus related problems can be found in Machado et al. (2011).

References

Blu, T., Unser, M.: Fractional spline wavelet software for MATLAB. http://bigwww.epfl.ch/demo/
    fractsplines/matlab.html (1999)
Bultheel, A.: Digital watermarking of images in the fractional Fourier domain. Report TW 497
    http://ftp.cs.kuleuven.ac.be/publicaties/rapporten/tw/TW497.pdf (2007)
Bultheel, A., Martínez Sulbaran, H.E.: Computation of the fractional Fourier transform. Appl.
    Comput. Harmonic Anal. **16**(3), 182–202 (2004)
Chen, C.: MATLAB File Exchange http://www.mathworks.com/matlabcentral/fileexchange/
    19148-hurst-parameter-estimat (2008)
Chen, Y., Vinagre, B.M.: A new IIR-type digital fractional order differentiator. Signal Process.
    83(11), 2359–2365 (2003). doi:10.1016/s0165-1684(03)00188-9
Chen, Y., Vinagre, B.M., Podlubny, I.: Continued fraction expansion approaches to discretizing
    fractional order derivatives—an expository review. Nonlinear Dyn. **38**(1), 155–170 (2004).
    doi:10.1007/s11071-004-3752-x
Fatichi, S.: MATLAB File Exchange http://www.mathworks.com/matlabcentral/fileexchange/
    25611-arfima-simulations (2009)
Hongguang, S.: MATLAB File Exchange http://www.mathworks.com/matlabcentral/fileexchange/
    authors/82631 (2010)
Machado, J.T., Kiryakova, V., Mainardi, F.: Recent history of fractional calculus. Commu-
    nications in Nonlinear Science and Numerical Simulation **16**(3), 1140–1153 (2011).
    doi:10.1016/j.cnsns.2010.05.027
Monje, C.A., Chen, Y.Q., Vinagre, B.M., Xue, D., Feliu, V.: Fractional-order Systems and Controls:
    Fundamentals and Applications. Springer (2010)
Nolan, J.: http://academic2.american.edu/~jpnolan/stable/stable.html (2000)
Oustaloup, A., Melchior, P., Lanusse, P., Cois, O., Dancla, F.: The CRONE toolbox for Matlab.
    In: IEEE International Symposium on Computer-aided Control System Design, 2000, CACSD
    2000, pp. 190–195 (2000)
Podlubny, I.: http://www.mathworks.com/matlabcentral/fileexchange/authors/8245 (2009)
Sheng, H.: http://www.mathworks.com/matlabcentral/fileexchange/authors/82211 (2010)

Sierociuk, D.: Fractional order discrete state-space system simulink toolkit user guide. http://www.ee.pw.edu.pl/~dsieroci/fsst/fsst.pdf (2005)

Stoev, S., Taqqu, M.: Simulation methods for linear fractional stable motion and FARIMA using the fast Fourier transform. Fractals (1), 95–122 (2004)

Stoev, S., Taqqu, M.S., Park, C., Michailidis, G., Marron, J.S.: LASS: a tool for the local analysis of self-similarity. Comput. Stat. Data. Anal. **50**(9), 2447–2471 (2006). doi:10.1016/j.csda.2004.12.014

Taqqu, M.: http://math.bu.edu/people/murad/methods

Valerio, D.: Toolbox ninteger for MATLAB v. 2.3. http://web.ist.utl.pt/duarte.valerio/ninteger/ninteger.html (2005)

Veillette, M.: http://math.bu.edu/people/mveillet/html/alphastablepub.html (2009)

Veitch, D.: http://www.cubinlab.ee.unimelb.edu.au/~darryl/secondorder_code.html (2007)

Xue, D., Chen, Y.: Solving Applied Mathematical Problems with Matlab. Chapman & Hall, CRC (2009)

Xue, D., Chen, Y.Q., Atherton, D.P., Industrial, S.f., Mathematics, A.: Linear feedback control: analysis and design with MATLAB. Society for Industrial and Applied Mathematics, Philadelphia. Taylor & Francis Group, Boca Raton (2007). ISBN:13:978-1-4200-8250-0

Yang Quan, C., Petras, I., Dingyu, X.: Fractional order control—a tutorial. In: American Control Conference, ACC '09, 10–12 June 2009, pp. 1397–1411